Student Solutions Manual
for Rice and Strange's

Algebra and Trigonometry
with Applications

FOURTH EDITION

Carroll M. Schleppi
University of Dayton

Brooks/Cole Publishing Company
Pacific Grove, California

Brooks/Cole Publishing Company
A Division of Wadsworth, Inc.

Printed in the United States of America

10   9   8   7   6   5   4   3   2   1

ISBN 0-534-10196-8

Sponsoring Editors: Faith Stoddard and Sue Ewing
Production: Dorothy Bell
Cover Design: Flora Pomeroy

An Introduction for Students

Every odd-numbered problem with an answer that requires some work to reach a solution has been solved in this manual. The method used in each case matches the method used in that section of the text. Some simple steps have been excluded.

Use this manual to check your solutions or to assist you when you are having difficulty solving a problem. Be aware that often several correct methods for doing a problem may be used. The method you select may be different from the one included in the manual, but also correct. It is unwise to use the manual as a substitute for solving the problems in the text.

If you have difficulty following any of the solutions provided, please feel free to contact me c/o Brooks/Cole Publishing Company, 511 Forest Lodge Road, Pacific Grove, California 93950.

Carroll M. Schleppi

University of Dayton

Contents

13 Sequences, Probability, and Mathematical Induction   229

14 The Conic Sections   245

Chapter 1   Basic Algebra

21. $-2 - (7-10)(5) = -2 - (-3)(5) = -2 + 15 = 13$

23. $-\big[2 + 3(6-8)\big]\big[4 - (5+2)\big] = -\big[2 + 3(-2)\big]\big[4-7\big] = -\big[2 + 3(-2)\big]\big[-3\big] = -\big[-4\big]\big[-3\big]$
     $= -12$

25. $\big[-(-2) - (4-6)\big]\big[3 + (7-3)\big] = \big[2-(-2)\big]\big[3 + 4\big] = [4][7] = 28$

27. $-2\big[7.1 + (1.2-2.8) - 6(2.5-3.2)\big] = -2\big[7.1 - 1.6 - 6(-.7)\big] = -2\big[5.5+4.2\big] = -2\big[9.7\big]$
     $= -19.4$

51. $|6-9| = |-3| = 3$

53. $5 + |-9| = 5 + 9 = 14$

55. $3 + |-3| = 3 + 3 = 6$

57. $1 - |3| = 1 - 3 = -2$

59. $d = |-3 - 1| = 4$

61. $d = |-5 - (-1)| = |-5 + 1| = 4$

63. $d = |-0.1 - 7| = 7.1$

65. $d = |-6 - 0| = 6$

67. $d = |-0.125 - 0.176| = 0.301$

69. $|a+b| \neq |a| + |b|$

   Try $a = 5,\ b = -3$

   $|5-3| = 2 \neq |5| + |-3| = 8$

   $|a+b| = |a| + |b|,$ when $a \geq 0$ and $b \geq 0$
   or $a < 0$ and $b < 0$

21. $\dfrac{7x^2 y^5}{27xy} = \dfrac{7}{27}xy^4$

23. $\dfrac{(x^2)^6}{(y^5)^6} = \dfrac{x^{12}}{y^{30}}$

29. $(x^{-2}y^5)^4 = \left(\dfrac{y^5}{x^2}\right)^4 = \dfrac{y^{20}}{x^8}$

31. $\left(\dfrac{y}{2x}\right)^3 = \dfrac{y^3}{8x^3}$

33. $\left(\dfrac{a^{-3}b^7}{a^2 b^4}\right)^{-2} = \left(\dfrac{b^3}{a^5}\right)^{-2} = \left(\dfrac{a^5}{b^3}\right)^2 = \dfrac{a^{10}}{b^6}$

35. $\left(\dfrac{x^3 y^{-6}}{x^7 y^{-4}}\right)^{-2} = \left(\dfrac{1}{x^4 y^2}\right)^{-2} = \left(x^4 y^2\right)^2 = x^8 y^4$

39. $x^0 + (2y)^0 = 1 + 1 = 2$

57. $(x+y)^{-1} \overset{?}{=} x^{-1} + y^{-1}$

Consider $x = 2$, $y = 3$

$(2+3)^{-1} = \frac{1}{5}$

$2^{-1} + 3^{-1} = \frac{1}{2} + \frac{1}{3} = \frac{5}{6}$

Therefore

$(x+y)^{-1} \neq x^{-1} + y^{-1}$

59. $(x^2)^3 \overset{?}{=} x^{2^3}$

Consider $x = 2$

$(2^2)^3 = 4^3 = 64$

$2^{2^3} = 2^8 = 256$

So $(x^2)^3 \neq x^{2^3}$

Section 1.3   Radicals and Fractional Exponents

1. $\sqrt{49} = \sqrt{7^2} = 7$

3. $\sqrt[3]{125} = \sqrt[3]{5^3} = 5$

5. $\sqrt[5]{-32} = \sqrt[5]{(-2)^5} = -2$

7. $\sqrt[7]{5^{14}} = \sqrt[7]{(5^2)^7} = 5^2 = 25$

9. $5\sqrt{243} = 5\sqrt{9^2 \cdot 3} = 5 \cdot 9\sqrt{3} = 45\sqrt{3}$

11. $\sqrt{a^4 b^2} = \sqrt{(a^2)^2 b^2} = a^2 b$

13. $\sqrt{63 a^5 b^8} = \sqrt{7}\sqrt{3^2}\sqrt{(a^2)^2 a}\sqrt{(b^4)^2} = 3a^2 b \bullet \sqrt[4]{7a}$

15. $\sqrt[4]{2 x^3 z^{12}} = \sqrt[4]{2}\,\sqrt[4]{x^3}\,\sqrt[4]{(z^3)^4} = z^3 \sqrt[4]{2x^3}$

17. $2m\sqrt{m^5 n^3} = 2m\sqrt{(m^2)^2 m}\sqrt{n^2 n} = 2m \cdot m^2 n\sqrt{mn} = 2m^3 n\sqrt{mn}$

19. $\sqrt{\dfrac{x^7}{y^4}} = \dfrac{\sqrt{(x^3)^2 x}}{\sqrt{(y^2)^2}} = \dfrac{x^3 \sqrt{x}}{y^2}$

21. $\dfrac{2}{\sqrt{5}} \cdot \dfrac{\sqrt{5}}{\sqrt{5}} = \dfrac{2\sqrt{5}}{5}$

23. $\dfrac{1}{\sqrt{a}} \cdot \dfrac{\sqrt{a}}{\sqrt{a}} = \dfrac{\sqrt{a}}{a}$

25. $\sqrt{5}\sqrt{20} = \sqrt{100} = 10$

27. $\sqrt[3]{2a}\,\sqrt[3]{4a^2} = \sqrt[3]{8a^3} = \sqrt[3]{2^3}\,\sqrt[3]{a^3} = 2a$

35. $\sqrt{x\sqrt{x\,\sqrt[3]{x}}} = \left(x\left(x(x^{1/3}) \right)^{1/2} \right)^{1/2} = \left(x\left(x^{4/3} \right)^{1/2} \right)^{1/2} = \left(x \cdot x^{2/3} \right)^{1/2} = \left(x^{5/3} \right)^{1/2} = x^{5/6}$

39. $\left(a^{3/2} \right)^{1/4} = a^{3/8} = \sqrt[8]{a^3}$

45. $81^{3/4} = \left(81^{1/4} \right)^3 = 3^3 = 27$

47. $(32)^{-3/5} = \left(32^{1/5} \right)^{-3} = 2^{-3} = \frac{1}{8}$

49. $x^{1/2} x^{2/3} = x^{1/2 + 2/3} = x^{7/6}$

51. $y^{4/3} y^{-1/2} = y^{(4/3)(-1/2)} = y^{5/6}$

53. $(125 a^3 b^9)^{1/3} = 125^{1/3} ab^3 = 5ab^3$

55. $\left(9x^{2/3} \right)^{1/2} = 9^{1/2} x^{1/3} = 3x^{1/3}$

57. $\sqrt[4]{y}\,\sqrt[3]{y} = y^{1/4} y^{1/3} = y^{1/4 + 1/3} = y^{7/12} = \sqrt[12]{y^7}$

59. $\sqrt[3]{5}\,\sqrt[4]{4} = 5^{1/3} 4^{1/4} = 5^{4/12} 4^{3/12} = (5^4 4^3)^{1/12} = \sqrt[12]{5^4 4^3}$

61. $\sqrt{x}\,\sqrt[3]{(x+y)} = x^{1/2}(x+y)^{1/3} = x^{3/16}(x+y)^{2/6} = \left(x^3(x+y)^2 \right)^{1/6} = \sqrt[6]{x^3(x+y)^2}$

63. $\sqrt{\sqrt[5]{x^4}} = \left(x^{4/5}\right)^{1/2} = x^{2/5} = \sqrt[5]{x^2}$ 65. $27^{1/6} = \left(27^{1/3}\right)^{1/2} = \sqrt{3}$

67. $\sqrt[4]{(a+1)^2} = (a+1)^{2/4} = \sqrt{a+1}$ 69. $\sqrt[6]{(y+2)^3} = (y+2)^{3/6} = \sqrt{y+2}$

79. $v = kT^{1.2} = (2.3 \times 10^4)(13^{1.2}) = 4.994 \times 10^5$

81. $\left(a^{2/3}+b^{2/3}\right)^{3/2} = \left(1.3^{2/3}+2.5^{2/3}\right)^{3/2} = 5.28$

$\left(\sqrt{a}+\sqrt{b}\right)^2 = \left(\sqrt{1.3}+\sqrt{2.5}\right)^2 = 7.41$

Section 1.4   Multiplication of Algebraic Expressions

1. $x^2(x^3+2) = x^2 x^3 + x^2(2) = x^5 + 2x^2$

3. $x^3(x^2-5x) = x^3 x^2 + x^3(-5x) = x^5 - 5x^4$

5. $(x+2y)(x-2y) = x(x-2y) + 2y(x-2y) = x^2 - 2xy + 2xy - 4y^2 = x^2 - 4y^2$

7. $(x^2-1)(x^2+1) = x^2(x^2+1) - (x^2+1) = x^4 + x^2 - x^2 - 1 = x^4 - 1$

9. $\sqrt{x}(\sqrt{x}+\sqrt{y}) = \sqrt{x}\sqrt{x} + \sqrt{x}\sqrt{y} = x + \sqrt{xy}$

11. $(xy+1)(xy-1) = xy(xy-1) + (xy-1) = x^2 y^2 - xy + xy - 1 = x^2 y^2 - 1$

13. $(x+y^2)^3 = \left[(x+y^2)(x+y^2)\right](x+y^2)$

$= \left[x(x+y^2) + y^2(x+y^2)\right](x+y^2)$

$= \left[x^2+xy^2+xy^2+y^4\right](x+y^2)$

$= \left[x^2+2xy^2+y^4\right]x + \left[x^2+2xy^2+y^4\right]y^2$

$= x^3 + 2x^2 y^2 + xy^4 + x^2 y^2 + 2xy^4 + y^6$

$= x^3 + 3x^2 y^2 + 3xy^4 + y^6$

15. $(x+y)(x-y)(x^2-y)$

$= \left[x(x-y) + y(x-y)\right](x^2-y)$

$= \left[x^2-xy+xy-y^2\right](x^2-y)$

$= (x^2-y^2)x^2 + (x^2-y^2)(-y)$

$= x^4 - x^2 y^2 - x^2 y + y^3$

17. $(x+y+z)(x+y+z) = x(x+y+z) + y(x+y+z) + z(x+y+z)$

$= x^2 + xy + xz + xy + y^2 + yz + xz + yz + z^2$

$= x^2 + 2xy + 2xz + 2yz + y^2 + z^2$

19. $(3x+4y-z)(2x-3y) = 3x(2x-3y) + 4y(2x-3y) - z(2x-3y)$

$\qquad = 6x^2 - 9xy + 8xy - 12y^2 - 2xz + 3yz$

$\qquad = 6x^2 - 12y^2 - xy - 2xz + 3yz$

27. $(2x+5)^2 = (2x)^2 + 2(2x)5 + 5^2 = 4x^2 + 20x + 25$

29. $(\sqrt{a}-\sqrt{b})^2 = (\sqrt{a})^2 - 2\sqrt{a}\sqrt{b}+(\sqrt{b})^2 = a - 2\sqrt{ab} + b$

31. $(4x+5)(4x-5) = (4x)^2 - 5^2 = 16x^2 - 25$

35. $(5a+b)^3 = (5a)^3 + 3(5a)^2 b + 3(5a)b^2 + b^3 = 125a^3 + 75a^2 b + 15ab^2 + b^3$

37. $(2x-5)^3 = (2x)^3 - 3(2x)^2 5 + 3(2x)5^2 - 5^3 = 8x^3 - 60x^2 + 150x - 125$

43. $(\sqrt{C}+\sqrt{D})(C-\sqrt{CD}+D) = (\sqrt{C})^3 + (\sqrt{D})^3 = C\sqrt{C} + D\sqrt{D}$

45. $(x+5)(x+9) = x^2 + 9x + 5x + 45 = x^2 + 14x + 45$

47. $(2x-8)(3x+5) = 6x^2 + 10x - 24x - 40 = 6x^2 - 14x - 40$

49. $(2y+5)(y+2) = 2y^2 + 4y + 5y + 10 = 2y^2 + 9y + 10$

51. $(2x-3y)(x+4y) = 2x^2 + 8xy - 3xy - 12y^2 = 2x^2 + 5xy - 12y^2$

53. $(12x-5)(3x-8) = 36x^2 - 96x - 15x + 40 = 36x^2 - 111x + 40$

55. $\dfrac{(2+h)^3 - 8}{h} = \dfrac{8 + 3(4)h + 3(2)h^2 + h^3 - 8}{h} = \dfrac{\cancel{h}(12+6h+h^2)}{\cancel{h}} = 12 + 6h + h^2$

57. $(5t-3)(2t-1) = 10t^2 - 5t - 6t + 3 = 10t^2 - 11t + 3$

59. $4(2T+3)(3T+1) = 4(6T^2+2T+9T+3) = 4(6T^2+11T+3) = 24T^2 + 44T + 12$

Section 1.5   Factoring

5. $z^2 - 24z + 144 = z^2 - 24z + 12^2 = (z-12)^2$

7. $9x^2 + 6x + 1 = (3x)^2 + 6x + 1 = (3x+1)^2$

9. $\frac{1}{4}x^2 + x + 1 = (\frac{x}{2})^2 + x + 1 = (\frac{x}{2}+1)^2$

11. $x^2 - 169 = x^2 - 13^2 = (x-13)(x+13)$

13. $4m^2 - 64 = 4(m^2-16) = 4(m^2-4^2) = 4(m-4)(m+4)$

15. $x^2 - 7 = x^2 - (\sqrt{7})^2 = (x-\sqrt{7})(x+\sqrt{7})$

17. $\frac{1}{4}x^2 - \frac{1}{9} = \left(\frac{1}{2}x\right)^2 - \left(\frac{1}{3}\right)^2 = \left(\frac{1}{2}x-\frac{1}{3}\right)\left(\frac{1}{2}x+\frac{1}{3}\right)$

19. $9x^2 + 12x + 4 = (3x)^2 + 12x + 2^2 = (3x+2)^2$

21. $\frac{1}{9}x^2 + \frac{2}{9}x + \frac{1}{9} = \frac{1}{9}(x^2+2x+1) = \frac{1}{9}(x+1)^2$

39. $8x^2 - 20xy + 8y^2 = 4(2x^2-5xy+2y^2) = 4(2x-y)(x-2y)$

41. $x^3 + 64 = x^3 + 4^3 = (x+4)(x^2-4x+16)$

43. $8x^3 + 1 = (2x)^3 + 1 = (2x+1)(4x^2-2x+1)$

45. $a^3 + \frac{1}{8} = a^3 + \left(\frac{1}{2}\right)^3 = \left(a+\frac{1}{2}\right)\left(a^2-\frac{a}{2}+\frac{1}{4}\right)$

47. $y^3 - 9y^2 + 27y - 27 = y^3 - 9y^2 + 27y - 3^3 = (y-3)^3$

49. $8x^3 + 12x^2 + 6x + 1 = (2x)^3 + 12x^2 + 6x + 1 = (2x+1)^3$

51. $a^2 + a^3b + ab^3 + b^2 = a^2(1+ab) + b^2(ab+1) = (1+ab)(a^2+b^2)$

53. $z^2 - w^2 - 3z + 3w = (z-w)(z+w) - 3(z-w) = (z-w)(z+w-3)$

55. $a^2 - c^2 + 2cd - d^2 = a^2 - (c^2-2cd+d^2) = a^2 - (c-d)^2$

$\quad = \left[a-(c-d)\right]\left[a+(c-d)\right] = (a-c+d)(a+c-d)$

57. $x^2+4x+1 = x^2 + 4x + 4 - 4 + 1$

$\qquad = x^2 + 4x + 4 - 3$

$\qquad = (x+2)^2 - 3$

$\qquad = (x+2-\sqrt{3})(x+2+\sqrt{3})$

59. $a^2-8a-5 = a^2 - 8a + 16 - 16 - 5$

$\qquad = (a-4)^2 - 21 = (a-4-\sqrt{21})(a-4+\sqrt{21})$

61. $x^2+3x+\frac{3}{4} = x^2 + 3x + \frac{9}{4} - \frac{9}{4} + \frac{3}{4}$

$\qquad = \left(x+\frac{3}{2}\right)^2 - \frac{3}{2} = \left(x+\frac{3}{2}-\sqrt{\frac{3}{2}}\right)\left(x+\frac{3}{2}+\sqrt{\frac{3}{2}}\right)$

63. $4y^2-8y-8 = 4(y^2-2y+1) - 4 - 8$

$\qquad = 4(y-1)^2 - 12 = \left[2(y-1)\right]^2 - (\sqrt{12})^2$

$\qquad = \left[2(y-1)-\sqrt{12}\right]\left[2(y-1)+\sqrt{12}\right]$

$\qquad = \left[2(y-1)-2\sqrt{3}\right]\left[2(y-1)+2\sqrt{3}\right]$

$\qquad = 4(y-1-\sqrt{3})(y-1+\sqrt{3})$

65. $\quad 2b^2+6b-3 = 2\left(b^2+3b+\frac{9}{4}\right) - \frac{9}{2} - 3$

$\qquad = 2\left(b+\frac{3}{2}\right)^2 - \frac{15}{2} = 2\left(b+\frac{3}{2}\right)^2 - \frac{15(2)}{4}$

$\qquad = 2\left(b+\frac{3}{2}-\frac{\sqrt{15}}{2}\right)\left(b+\frac{3}{2}+\frac{\sqrt{15}}{2}\right)$

67. $x^2+Lx-2L^2 = \left(x^2+Lx+\frac{L^2}{4}\right) - \frac{L^2}{4} - 2L^2$

$$= \left(x+\frac{L}{2}\right)^2 - \frac{9L^2}{4} = \left(x+\frac{L}{2}-\frac{3L}{2}\right)\left(x+\frac{L}{2}+\frac{3L}{2}\right) = (x-L)(x+2L)$$

69. $5R - 5r + r(R-r) = 5(R-r) + r(R-r) = (5+r)(R-r)$

71. $\sqrt{5+2\sqrt{6}} = \sqrt{2+2\sqrt{6}+3} = \sqrt{(\sqrt{2})^2+2\sqrt{6}+(\sqrt{3})^2} = \sqrt{(\sqrt{2}+\sqrt{3})^2} = \sqrt{2} + \sqrt{3}$

73. $\sqrt{71-16\sqrt{7}} = \sqrt{64-16\sqrt{7}+7} = \sqrt{8^2-16\sqrt{7}+(\sqrt{7})^2} = \sqrt{(8-\sqrt{7})^2} = 8 - \sqrt{7}$

75. $\sqrt{16+6\sqrt{7}} = \sqrt{9+6\sqrt{7}+7} = \sqrt{3^2+6\sqrt{7}+(\sqrt{7})^2} = \sqrt{(3+\sqrt{7})^2} = 3 + \sqrt{7}$

77. $\sqrt[3]{7+5\sqrt{2}} = \sqrt[3]{6+1+5\sqrt{2}} = \sqrt[3]{1+3\sqrt{2}+3(2)+2\sqrt{2}}$

$$= \sqrt[3]{1+3\sqrt{2}+3(\sqrt{2})^2+(\sqrt{2})^3} = \sqrt[3]{(1+\sqrt{2})^3} = 1 + \sqrt{2}$$

Section 1.6   Fractional Expressions

3. $\dfrac{3x}{4y^3} \div \dfrac{9x^4}{2y} = \dfrac{3x}{4y^3} \cdot \dfrac{2y}{9x^4} = \dfrac{xy}{6x^4y^3} = \dfrac{1}{6x^3y^2}$

5. $\dfrac{9b^2}{4c^3} \cdot \dfrac{24ac^2}{18b^5} \div \dfrac{c}{2a^3b} = \dfrac{9b^2}{4c^3} \cdot \dfrac{24ac^2}{18b^5} \cdot \dfrac{2a^3b}{c} = \dfrac{6a^4b^3c^2}{b^5c^4}$

$$= \dfrac{6a^4}{b^2c^2}$$

7. $\dfrac{(x+3)(x-2)}{(x+3)(x-3)} \cdot \dfrac{(x-3)^2}{2(x-2)} = \dfrac{x-3}{2}$

9. $\dfrac{4xy^2}{(2x-3)(2x+3)} \cdot \dfrac{3(2x+3)}{10x^2y} = \dfrac{6xy^2}{5x^2y(2x-3)} = \dfrac{6y}{5x(2x-3)}$

11. $\dfrac{(x-y)}{(x+y)} \cdot \dfrac{x(x+y)}{y(x-y)} \cdot \dfrac{y(x-2)}{x(x-1)} = \dfrac{x(x-2)}{y(x-1)}$

13. $\dfrac{(2y+1)(y+5)}{(2y+3)(y+2)} \cdot \dfrac{(2y+3)(y-1)}{(y+5)(y-1)} = \dfrac{2y+1}{y+2}$

15. $\dfrac{(2x+3)\cancel{(x-1)}}{\cancel{(3x-2)}\cancel{(x+4)}} \cdot \dfrac{x\cancel{(3x-2)}}{\cancel{(x-1)}\cancel{(x+1)}} \cdot \dfrac{\cancel{(x+4)}\cancel{(x+1)}}{(x+6)} = \dfrac{x(2x+3)}{x+6}$

17. $\dfrac{s^2-2s-3}{s-2} \cdot \dfrac{(s+1)(s-2)}{(s-3)(s+3)} = \dfrac{\cancel{(s-3)}(s+1)}{\cancel{s-2}} \cdot \dfrac{(s+1)\cancel{(s-2)}}{\cancel{(s-3)}(s+3)} = \dfrac{(s+1)^2}{s+3}$

19. $\dfrac{a(x+y)+b(x+y)}{a(x-y)+b(x-y)} \cdot \dfrac{x-y}{x+y} = \dfrac{\cancel{(a+b)}\cancel{(x+y)}}{\cancel{(a+b)}\cancel{(x-y)}} \cdot \dfrac{\cancel{(x-y)}}{\cancel{(x+y)}} = 1$

21. $\dfrac{(2y+1)\cancel{(y-3)}}{\cancel{y-4}} \cdot \dfrac{\cancel{(y-4)}(y+4)(3-y)}{(y-3)^2} = \dfrac{(2y+1)(y+4)(-1)\cancel{(y-3)}}{\cancel{(y-3)}} = -(2y+1)(y+4)$

23. $\dfrac{\cancel{(a+200)}(a+100)}{(40-a)(40+a)} \cdot \dfrac{(a-40)(a+400)}{2\cancel{(a+200)}} = \dfrac{(a+100)\cancel{(a-40)}(a+400)}{(-1)\cancel{(a-40)}(a+40)(2)} = \dfrac{-(a+100)(a+400)}{2(a+40)}$

25. $\dfrac{2(ac)}{b(ac)} + \dfrac{1(ab)}{c(ab)} - \dfrac{3(bc)}{a(bc)} = \dfrac{2ac+ab-3bc}{abc}$

27. $\dfrac{2}{3s} - \dfrac{1}{t} + \dfrac{1}{3st} = \dfrac{2(t)}{3st} - \dfrac{1(3s)}{3st} + \dfrac{1}{3st} = \dfrac{2t-3s+1}{3st}$

29. $\dfrac{3}{xy^2} + \dfrac{2(y^2)}{xy^2} - \dfrac{3(xy)}{xy^2} = \dfrac{3+2y^2-3xy}{xy^2}$

31. $\dfrac{3(x-3)}{(x+1)(x-3)} + \dfrac{2(x+1)}{(x+1)(x-3)} = \dfrac{3(x-3)+2(x+1)}{(x+1)(x-3)} = \dfrac{3x-9+2x+2}{(x+1)(x-3)} = \dfrac{5x-7}{(x+1)(x-3)}$

33. $\dfrac{x}{x-5} + \dfrac{2}{x+3} = \dfrac{x(x+3)+2(x-5)}{(x-5)(x+3)} = \dfrac{x^2+3x+2x-10}{(x-5)(x+3)} = \dfrac{x^2+5x-10}{(x-5)(x+3)}$

35. $\dfrac{3(x+2)-2x-1}{x(x+2)} = \dfrac{3x+6-2x-1}{x(x+2)} = \dfrac{x+5}{x(x+2)}$

37. $\dfrac{2(2a+3)+3a-2}{a(2a+3)} = \dfrac{4a+6+3a-2}{a(2a+3)} = \dfrac{7a+4}{a(2a+3)}$

39. $\dfrac{4(3x+4)-(9x-16)}{3x+4} = \dfrac{12x+16-9x+16}{3x+4} = \dfrac{3x+32}{3x+4}$

41. $\dfrac{2(p+1)^2+2p-p(p+1)}{p(p+1)^2} = \dfrac{2(p^2+2p+1)+2p-p^2-p}{p(p+1)^2}$

$= \dfrac{2p^2+4p+2+p-p^2}{p(p+1)^2} = \dfrac{p^2+5p+2}{p(p+1)^2}$

43. $\dfrac{(x+1)+(x-2)-(x+3)}{(x+1)(x-2)(x+3)} = \dfrac{x-4}{(x+1)(x-2)(x+3)}$

45. $\dfrac{5x+6}{(x+3)(x-2)} + \dfrac{x+1}{(x-2)^2} - \dfrac{\frac{1}{x^2}}{x^2(x-2)}$

$$= \dfrac{(5x+6)(x-2) + (x+1)(x+3) - (x+3)(x-2)}{(x+3)(x-2)^2}$$

$$= \dfrac{(5x^2-4x-12) + (x^2+4x+3) - (x^2+x-6)}{(x+3)(x-2)^2} = \dfrac{5x^2-x-3}{(x-2)^2(x+3)}$$

47. $\dfrac{x+3}{\sqrt{x+2}} + \sqrt{x-2} = \dfrac{x+3+(\sqrt{x-2})^2}{\sqrt{x-2}} = \dfrac{2x+1}{\sqrt{x+2}} \cdot \dfrac{\sqrt{x+2}}{\sqrt{x-2}} = \dfrac{(2x+1)\sqrt{x-2}}{x-2}$

49. $\dfrac{\sqrt{8}}{\sqrt{3}+\sqrt{5}} \cdot \dfrac{\sqrt{3}-\sqrt{5}}{\sqrt{3}-\sqrt{5}} = \dfrac{2\sqrt{2}(\sqrt{3}-\sqrt{5})}{3-5} = \dfrac{2\sqrt{2}(\sqrt{3}-\sqrt{5})}{-2}$

$$= \sqrt{2}(\sqrt{5}-\sqrt{3})$$

51. $\dfrac{\sqrt{x}-\sqrt{y}}{\sqrt{x}+\sqrt{y}} \cdot \dfrac{\sqrt{x}-\sqrt{y}}{\sqrt{x}-\sqrt{y}} = \dfrac{(\sqrt{x}-\sqrt{y})^2}{x-y} = \dfrac{x-2\sqrt{xy}+y}{x-y}$

53. $\dfrac{\frac{x^2-1}{x}}{\frac{x-1}{x}} = \dfrac{x^2-1}{\cancel{x}} \cdot \dfrac{\cancel{x}}{x-1} = \dfrac{\cancel{(x-1)}(x+1)}{\cancel{x-1}} = x+1$

55. $\dfrac{\frac{a^2-b^2}{ab}}{\frac{a+b}{a}} = \dfrac{(a-b)\cancel{(a+b)}}{\cancel{a}b} \cdot \dfrac{\cancel{a}}{\cancel{a+b}} = \dfrac{a-b}{b}$

57. $\dfrac{1-\frac{1}{x^2}}{\frac{1}{x}-1} = \dfrac{\frac{x^2-1}{x^2}}{\frac{1-x}{x}} = \dfrac{\cancel{(x-1)}(x+1)}{x^2} \cdot \dfrac{\cancel{x}}{\underset{-1}{\cancel{-(x-1)}}} = \dfrac{x+1}{-x} = \dfrac{-(x+1)}{x}$

59. $\dfrac{\frac{c^2+d^2}{cd}}{\frac{d+c}{cd}} = \dfrac{c^2+d^2}{cd} \cdot \dfrac{cd}{c+d} = \dfrac{c^2+d^2}{c+d}$

61. $\dfrac{\frac{2}{x-1}-x-2}{1-x+\frac{2}{x+2}} = \dfrac{\frac{2-x(x-1)-2(x-1)}{x-1}}{\frac{x+2-x(x+2)+2}{x+2}} = \dfrac{-x^2-x+4}{x-1} \cdot \dfrac{x+2}{-x^2-x+4} = \dfrac{x+2}{x-1}$

63. $\dfrac{\dfrac{x(x-1) + (x-4)}{(x-4)(x-1)}}{\dfrac{x(x-3) + 2(x-1)}{(x-1)(x-3)}} = \dfrac{x^2 - 4}{(x-4)(x-1)} \cdot \dfrac{(x-1)(x-3)}{x^2 - x - 2}$

$= \dfrac{(x+2)(x-2)(x-3)}{(x-4)(x-2)(x+1)} = \dfrac{(x+2)(x-3)}{(x-4)(x+1)}$

65. $\dfrac{\dfrac{x - y + y}{x - y}}{x - \dfrac{\dfrac{y}{x + y - x}}{x + y}} = \dfrac{\dfrac{x}{x - y}}{x - y \cdot \dfrac{(x + y)}{y}} = \dfrac{x}{x - y} \cdot \dfrac{1}{-y} = \dfrac{x}{-y(x-y)} = \dfrac{x}{y(y-x)}$

67. $\dfrac{4t^3 + 2(3t^2) - 5(6)}{12t^3} = \dfrac{2(2t^3 + 3t^2 - 15)}{2(6t^3)} = \dfrac{2t^3 + 3t^2 - 15}{6t^3}$

69. $\dfrac{3}{s(s+1)} + \dfrac{2}{(s+3)(s+1)} = \dfrac{3(s+3) + 2s}{s(s+3)(s+1)} = \dfrac{5s + 9}{s(s+1)(s+3)}$

Section 1.7   Common Errors

5. $\dfrac{1}{2} + \dfrac{3}{4} = \dfrac{2 + 3}{4} = \dfrac{5}{4}$ 7. $\dfrac{x}{x} + \dfrac{2}{x} = 1 + \dfrac{2}{x}$ 9. $\left(\dfrac{1}{x} + \dfrac{1}{y}\right)^{-1} = \left(\dfrac{x+y}{xy}\right)^{-1} = \dfrac{xy}{x + y}$

Chapter 1   Review

15. $|5-9| = |-4| = 4$

17. $|-10 - (-4)| = |-10 + 4| = |-6| = 6$

19. $-22 - |-39| = -22 - 39 = \ \ 61$ 21. $d = |5-14| - |-9| = 9$

23. $d = |-2 - 0| = |-2| = 2$ 25. $s = |-10 - (-7)| = |-3| = 3$

27. $a - 2a - b + c = -a - b + c$

29. $5x - 6\big[2x - 3x - 7y - y\big] = 5x - 6\big[-x - 8y\big] = 5x + 6x + 48y = 11x - 48y$

33. $x(3x+7y) - 2y(3x+7y) = 3x^2 + 7xy - 6xy - 14y^2 = 3x^2 + xy - 14y^2$

35. $2y(x-3y+5) + 3(x-3y+5) = 2xy - 6y^2 + 10y + 3x - 9xy + 15$
$= 2xy - 6y^2 + y + 3x + 15$

39. $2^3 + 3(2^2)3y + 3(2)(3y)^2 + (3y)^3 = 8 + 36y + 54y^2 + 27y^3$

59. $a(b-x) - 3(b-x) = (a-3)(b-x)$

61. $\dfrac{2x}{3xy} \cdot \dfrac{5x^2 y^3}{4xy^4} \cdot \dfrac{3x^3 y^2}{15x^4 y} = \dfrac{x^6 y^5}{6x^6 y^6} = \dfrac{1}{6y}$

63. $\dfrac{(a+3)^{\cancel{2}}}{(a-b)\cancel{(a+b)}} \cdot \dfrac{a\cancel{(a+b)}}{b\cancel{(a+3)}} = \dfrac{a(a+3)}{b(a-b)}$

65. $\dfrac{\cancel{(r-2)^2}}{\cancel{(r-3)(r+3)}} \cdot \dfrac{\cancel{(r-3)}(r+1)}{\cancel{(r-2)}(r+2)} \cdot \dfrac{r\cancel{(r+3)}}{\cancel{(r-2)}} = \dfrac{r(r+1)}{r+2}$

67. $\dfrac{2a}{ab} - \dfrac{3a}{ab} + \dfrac{1}{ab} = \dfrac{2b - 3a + 1}{ab}$

69. $\dfrac{2x(x+3) + 3x - (x+3)}{x(x+3)} = \dfrac{2x^2 + 8x - 3}{x(x+3)}$

71. $\dfrac{a}{a-2} + \dfrac{1}{(a-5)^2} = \dfrac{a(a-5)^2 - (a-2)}{(a-2)(a-5)^2} = \dfrac{a^3 - 10a^2 + 25a - a + 2}{(a-2)(a-5)^2} = \dfrac{a^3 - 10a^2 + 24a + 2}{(a-2)(a-5)^2}$

73. $\dfrac{\dfrac{x - y - x - y}{(x+y)(x-y)}}{\dfrac{x - y + x + y}{(x+y)(x-y)}} = \dfrac{-\cancel{2}y}{\cancel{(x+y)(x-y)}} \cdot \dfrac{\cancel{(x+y)(x-y)}}{\cancel{2}x} = \dfrac{-y}{x}$

77. $\left(\dfrac{x}{y^2} + 1\right)^{-1} = \left(\dfrac{x + y^2}{y^2}\right)^{-1} = \dfrac{y^2}{x + y^2}$

79. $\dfrac{x + \dfrac{1}{y}}{3 + \dfrac{1}{x^2} + \dfrac{1}{y^2}} = \dfrac{\dfrac{xy+1}{y}}{\dfrac{3x^2 y^2 + y^2 + x^2}{x^2 y^2}} = \dfrac{xy+1}{y} \cdot \dfrac{x^2 y^2}{3x^2 y + y^2 + x^2} = \dfrac{x^2 y(xy+1)}{3x^2 y^2 + y^2 + x^2}$

81. $\left(\dfrac{x^3 y^{-7}}{2x^5 y^{-2}}\right)^{-3} = \left(\dfrac{1}{2x^2 y^5}\right)^{-3} = (2x^2 y^5)^3 = 8x^6 y^{15}$

83. $\dfrac{\dfrac{1}{x} + \dfrac{1}{x^2}}{\dfrac{1}{x} - \dfrac{1}{x^2}} = \dfrac{\dfrac{x^2 + x}{x^2}}{\dfrac{x^2 - x}{x^2}} = \dfrac{\cancel{x}(x+1)}{\cancel{x^2}} \cdot \dfrac{\cancel{x^2}}{\cancel{x}(x-1)} = \dfrac{x + 1}{x + 1}$

85. $\dfrac{\dfrac{1}{a^2}}{1 - \dfrac{1}{a}} - \dfrac{\dfrac{1}{a}}{1 - \dfrac{1}{a^2}} = \dfrac{\dfrac{1}{a^2}}{\dfrac{a - 1}{a}} - \dfrac{\dfrac{1}{a}}{\dfrac{a^2 - 1}{a^2}}$

$= \dfrac{1}{a^2} \cdot \dfrac{a}{a - 1} - \dfrac{1}{a} \cdot \dfrac{a^2}{a^2 - 1} = \dfrac{1}{a(a-1)} - \dfrac{a}{(a-1)(a+1)}$

$= \dfrac{a + 1 - a^2}{a(a-1)(a+1)} = \dfrac{1 + a - a^2}{a(a-1)(a+1)}$

87. $\sqrt{16 \cdot 2} - \sqrt{9 \cdot 2} = 4\sqrt{2} - 3\sqrt{2} = \sqrt{2}$

89. $\sqrt{16 \cdot 3} + 5 \cdot 7 - 3\sqrt{3} = 4\sqrt{3} + 35 - 3\sqrt{3} = \sqrt{3} + 35$

91. $\dfrac{\sqrt{2}}{5+\sqrt{2}}\cdot\dfrac{5-\sqrt{2}}{5-\sqrt{2}} = \dfrac{5\sqrt{2}-2}{25-2} = \dfrac{5\sqrt{2}-2}{23} = \dfrac{\sqrt{2}(5-\sqrt{2})}{23}$

93. $\sqrt[5]{\sqrt{\sqrt[3]{9}}} = \left(\left(9^{1/3}\right)^{1/2}\right)^{1/5} = \left(9^{1/6}\right)^{1/5} = 9^{1/30} = \left(3^2\right)^{1/30} = 3^{1/15} = \sqrt[15]{3}$

95. $\sqrt{\dfrac{2}{c}+\dfrac{1}{d^2}} = \left(\dfrac{2d^2+c}{cd^2}\right)^{1/2} = \dfrac{1}{d}\left(\dfrac{2d^2+c}{c}\right)^{1/2} = \dfrac{1}{d}\left(\dfrac{c(2d^2+c)}{c^2}\right)^{1/2} = \dfrac{1}{cd}\left(2cd^2+c^2\right)^{1/2}$

$= \dfrac{1}{cd}\sqrt{2cd^2+c^2}$

97. $\dfrac{\sqrt{x}-\sqrt{y}}{\sqrt{x}+\sqrt{y}}\cdot\dfrac{\sqrt{x}-\sqrt{y}}{\sqrt{x}-\sqrt{y}} = \dfrac{(\sqrt{x}-\sqrt{y})^2}{x-y} = \dfrac{x-2\sqrt{xy}+y}{x-y}$

99. $\left[a^4(bc^3)^{-1}\right]^{1/3} = a^{4/3}(bc^3)^{-1/3} = \dfrac{\sqrt[3]{a^3\,a}}{\sqrt[3]{bc^3}} = \dfrac{a\cdot\sqrt[3]{a}}{c\cdot\sqrt[3]{b}}\cdot\dfrac{\sqrt[3]{b^2}}{\sqrt[3]{b^2}}$

$= \dfrac{a\cdot\sqrt[3]{ab^2}}{bc}$

101. $a^{1/3}a^{1/2} + a^{1/3}b + a^{1/2}b^{1/2} + bb^{1/2} = a^{5/6} + a^{1/3}b + a^{1/2}b^{1/2} + b^{3/2}$

103. $\left(x^{1/2}-\dfrac{1}{x^{1/2}}\right)^{-1} = \left(\dfrac{x-1}{x^{1/2}}\right)^{-1} = \dfrac{x^{1/2}}{x-1}$

105. $4x^{-1} = \dfrac{4}{x}$

107. $\left(\dfrac{1}{a^{1/3}}-2b^{1/3}\right)^2 = \left(\dfrac{1-2a^{1/3}b^{1/3}}{a^{1/3}}\right)^2 = \dfrac{1}{a^{2/3}}\left(1-4a^{1/3}b^{1/3}+4a^{2/3}b^{2/3}\right)$

$= \dfrac{1}{a^{2/3}}\left(1-4(ab)^{1/3}+4(ab)^{2/3}\right)$

109. $\left(ax^{1/2}+bx^{1/4}+c\right)\left(ax^{1/2}+bx^{1/4}+c\right)$

$= a^2x + abx^{3/4} + acx^{1/2} + abx^{3/4} + b^2x^{1/2} + bcx^{1/4} + cax^{1/2} + cbx^{1/4} + c^2$

$= a^2x + b^2x^{1/2} + c^2 + 2abx^{3/4} + 2acx^{1/2} + 2bcx^{1/4}$

111. $3 + \dfrac{5}{x+1} - \dfrac{x}{(x-3)(x+1)} = \dfrac{3(x^2-2x-3)+5(x-3)-x}{(x-3)(x+1)} = \dfrac{3x^2-2x-24}{(x-3)(x+1)}$

113. $F = (25-2t)(25+23t-2t^2)^{-1} = \dfrac{25-2t}{25+23t-2t^2} = \dfrac{\cancel{25-2t}}{\cancel{(25-2t)}(1+t)} = \dfrac{1}{1+t}$

Chapter 2   Equations and Inequalities

Section 2.1   Linear Equations

1. $2x+5 = 2$

$\quad 2x = -3$

$\qquad x = -\frac{3}{2}$

3. $\quad 6x-3-x = 4-x-3$

$\qquad 5x-3 = -x+1$

$\qquad\quad 6x = 4$

$\qquad\quad\; x = \frac{2}{3}$

5. $3x+5 = 4x-8$

$\quad 13 = x$

7. $\qquad \frac{4a}{3}-5a+2 = \frac{a}{2}-1$

$\quad 6\left(\frac{4a}{2}-5a+2\right) = \left(\frac{a}{2}-1\right)6$

$\qquad 8a-30a+12 = 3a-6$

$\qquad\qquad -25a = -18$

$\qquad\qquad\quad a = \frac{18}{25}$

9. $\frac{2x+1}{3}+16 = 3x$

$\quad 2x+1+48 = 9x$

$\qquad -7x = -49$

$\qquad\quad x = 7$

11. $\frac{5}{c} = 10$

$\quad 5 = 10c$

$\quad \frac{1}{2} = c$

13. $\frac{3x}{x+5}-4 = 0$

$\quad 3x-4(x+5) = 0$

$\qquad -x-20 = 0$

$\qquad\quad x = -20$

15. $\frac{3}{6y+2} = \frac{4}{7y+3}$

$\quad 3(7y+3) = 4(6y+2)$

$\quad 21y+9 = 24y+8$

$\qquad 1 = 3y$

$\qquad \frac{1}{3} = y$

17. $\qquad \frac{2}{x-2}-\frac{3}{x+5} = \frac{10}{(x-2)(x+5)}$

$\quad 2(x+5)-3(x-2) = 10$

$\quad 2x+10-3x+6 = 10$

$\qquad\qquad\quad x = 6$

$-\; 12 \;-$

19.
$$\frac{5}{5m-11} = \frac{3}{m-5} - \frac{4}{2m-3}$$

$$5(m-5)(2m-3) = 3(5m-11(2m-3)-4(5m-11)(m-5)$$

$$5(2m^2-13m+15) = 3(10m^2-37m+33)-4(5m^2-36m+55)$$

$$10m^2-65m+75 = 30m^2-111m+99-20m^2+144m-220$$

$$-65m+75 = 33m-121$$

$$-98m = -196$$

$$m = 2$$

21.
$$x^2+6x-7 = x^2+18x+81$$

$$-12x = 88$$

$$x = -\frac{88}{12} = -\frac{22}{3}$$

23.
$$2x^2+1 = 2x^2+5x-3$$

$$4 = 5x$$

$$x = \frac{4}{5}$$

25.
$$x^3+3x = 5+x^3$$

$$3x = 5$$

$$x = \frac{5}{3}$$

27.
$$\frac{5}{x+1} = \frac{3}{x+1} + \frac{2}{x-1}$$

$$5(x-1) = 3(x-1)+2(x+1)$$

$$5x-5 = 3x-3+2x+2$$

$$-5 = -1$$

No solution

29.

$x \geq 0$	$x < 0$
$x+2 = -2$	$-x+2 = -2$
$x = -4$	$x = 4$
Impossible	Impossible

31.

$x \geq 0$	$x < 0$
$3x-1 = x$	$3x-1 = -x$
$2x = 1$	$4x = 1$
$x = \frac{1}{2}$	$x = \frac{1}{4}$
	Impossible

33.

$2x-3 \geq 0$	$2x-3 < 0$
$2x \geq 3$	$2x < 3$
$x \geq \frac{3}{2}$	$x < \frac{3}{2}$

$2x-3 = -1$	$-(2x-3) = -1$
$2x = 2$	$-2x+3 = -1$
$x = 1$	$-2x = -4$
Impossible	$x = 2$
	Impossible

35.

$x+5 \geq 0$	$x+5 < 0$
$x \geq -5$	$x < -5$
$(x+5)-2x = 1+x$	$-(x+5)-2x = 1+x$
$-2x = -4$	$-4x = 6$
$x = 2$	$x = -\frac{6}{4} = -\frac{3}{2}$
	Impossible

37.

$5x-1 \geq 0$	$5x-1 < 0$
$5x \geq 1$	$5x < 1$
$x \geq \frac{1}{5}$	$x < \frac{1}{5}$
$(5x-1)+2 = 3x+39$	$-(5x-1)+2 = 3x+39$
$2x = 38$	$-8x = 36$
$x = 19$	$x = -\frac{36}{8} = -\frac{9}{2}$

Section 2.2   Formulas and Applications of Linear Equations

21.
$$v = at+b$$
$$v-b = at$$
$$\frac{v-b}{a} = t$$

23.
$$A = \tfrac{1}{2}h(a+b)$$
$$\frac{2a}{h} = a+b$$
$$\frac{2a}{h} - a = b$$

25.
$$L = L_0\left[1+\alpha(T_1-T_0)\right]$$
$$L = L_0+\alpha L_0 T_1 - \alpha L_0 T_0$$
$$L-L_0+\alpha L_0 T_0 = \alpha L_0 T_1$$
$$\frac{L-L_0+\alpha L_0 T_0}{\alpha L_0} = T_1$$
$$T_0+\frac{L-L_0}{\alpha L_0} = T_1$$
$$T_0+\tfrac{1}{\alpha}\left(\frac{L}{L_0}-1\right) = T_1$$

27.

$$D = \frac{nd}{n+2}$$

$$D(n+2) = nd$$

$$Dn + 2D = nd$$

$$Dn - nd = -2D$$

$$n(D-d) = -2D$$

$$n = \frac{-2D}{D-d} = \frac{-2D}{-(d-D)}$$

$$n = \frac{2D}{d-D}$$

29.

$$\frac{1}{R} = \frac{1}{R_1} + \frac{1}{R_2}$$

$$R_1 R_2 = RR_2 + RR_1$$

$$R_1 R_2 = R(R_2 + R_1)$$

$$\frac{R_1 R_2}{R_1 + R_2} = R$$

31.

$$x + y = 20$$

$$x = 2 + y$$

$$(2+y) + y = 20$$

$$2y = 18$$

$$y = 9$$

$$x = 11$$

33.

$$x, \quad x+2, \quad x+4$$

$$x + (x+2) + (x+4) = 312$$

$$3x + 6 = 312$$

$$3x = 306$$

$$x = 102$$

$$102, \quad 104, \quad 106$$

35.

	rate	time	distance
jogger 1	$\dfrac{1 \text{ mi}}{7 \text{ min}}$	t	$\dfrac{t}{7}$
jogger 2	$\dfrac{1 \text{ mi}}{6 \text{ min}}$	t	$\dfrac{t}{6}$

$$d_1 = d_2 - \frac{1}{10}$$

$$\frac{t}{7} = \frac{t}{6} - \frac{1}{10}$$

$$60\,t = 70t - 42$$

$$10t = 42$$

$$t = 4.2$$

$$d_2 = \frac{4.2}{6} = 0.7 \text{ mile}$$

37. city $= x$

    highway $= 500 - x$

$$\frac{x}{22} + \frac{500 - x}{28} = 20$$

$$28x + 22(500 - x) = 20(22)(28)$$

$$6x = 12{,}320 - 11{,}000 = 1320$$

$$x = \frac{1320}{6} = 220$$

39a. 75% wholesale price

    $x = 100 + 10(.75) = \$175$

39b. 75% retail price

$$x = 100 + .75x$$

$$.25x = 100$$

$$x = \$400$$

41.

	rate	time	distance
With wind	$400 + x$	1	$(400 + x)1$
Against wind	$400 - x$	2	$(400 - x)2$

$$400 + x = (400 - x)2$$

$$3x = 400$$

$$x = \frac{400}{3} = 133.3$$

43.

	rate	distance
Fast	170	$x + 2.5$
Slow	165	x

$$t = \frac{d}{r}$$

$$\frac{x + 2.5}{170} = \frac{x}{165}$$

$$165(x + 2.5) = 170x$$

$$2.5(165) = 5x$$

$$82.5 = x$$

$$t = \frac{82.5}{165} = .5$$

45. touchdown $= t$ field goals $= f$
    extra points $= e$ safeties $= s$

    $t = e = f = 2s$

    $t + e + f + s = 44$

    $6t + e + 3f + 2s = 44$

    $6t + t + 3t + 2\left(\dfrac{t}{2}\right) = 44$

    $11t = 44$

    $t = 4,\ e = 4$

    $f = 4,\ s = 2$

47. Girl $= x$
    Boy $= 20 - x$

    $100x = 150(20 - x)$

    $100x = 3000 - 150x$

    $250x = 3000$

    $x = 12$ feet from Girl

49.

	rate	time	distance
car	55	$x + .5$	$55(x + .5)$
truck	45	x	$45x$
total			206

$55(x + .5) + 45x = 206$

$100x + 27.5 = 206$

$x = 1.785$

Section 2.3   Quadratic Equations

1. $x(x - 2) = 0$

    $x = 0,\ 2$

3. $(x - 2)(x + 1) = 0$

    $x = -1,\ 2$

5. $x^2 - x - 6 = 0$

    $(x - 3)(x + 2) = 0$

    $x = 3,\ -2$

7. $(z + 5)(z - 2) = 0$

    $z = -5,\ 2$

9. $(2x - 3)(x + 4) = 0$

    $x = \dfrac{3}{2},\ -4$

11. $(3w + 1)(w - 2) = 0$

    $w = -\dfrac{1}{3},\ +2$

13. $2v(5v - 1) = 0$

    $v = 0,\ \dfrac{1}{5}$

15. $t^2 - 3t - 10 = 0$

    $(t - 5)(t + 2) = 0$

    $t = 5,\ -2$

17. $35x^2 + 2x - 24 = 0$

$(7x + 6)(5x - 4) = 0$

$x = -\frac{6}{7}, \frac{4}{5}$

19. $x = \dfrac{-4 \pm \sqrt{(-4)^2 - 4(3)(-4)}}{2(3)}$

$= \dfrac{-4 \pm \sqrt{64}}{6} = \dfrac{-4 + 8}{6}$ or $\dfrac{-4 - 8}{6} = \dfrac{2}{3}, -2$

21. $x = \dfrac{-2 \pm \sqrt{2^2 - 4(5)(-3)}}{2(5)} = \dfrac{-2 \pm \sqrt{64}}{10} = \dfrac{-2 \pm 8}{10} = -1, \dfrac{3}{5}$

23. $x = \dfrac{1 \pm \sqrt{(-1^2) - 4(2)(-2)}}{2(2)} = \dfrac{1 \pm \sqrt{17}}{4}$

25. $2x^2 - x - 4 = 0$

$x = \dfrac{1 \pm \sqrt{(-1)^2 - 4(2)(-4)}}{2(2)} = \dfrac{1 \pm \sqrt{33}}{4}$

27. $x = \dfrac{5 \pm \sqrt{(-5)^2 - 4(1)(5)}}{2} = \dfrac{5 \pm \sqrt{5}}{2}$

29. $x = \dfrac{-12 \pm \sqrt{(12)^2 - 4(5)(6)}}{2(5)} = \dfrac{-12 \pm \sqrt{24}}{10} = \dfrac{-12 \pm 2\sqrt{6}}{10} = \dfrac{2(-6 \pm \sqrt{6})}{10} = \dfrac{-6 \pm \sqrt{6}}{5}$

31. $x = \dfrac{2 \pm \sqrt{(-2)^2 - 4(3)(2)}}{2(3)} = \dfrac{2 \pm \sqrt{-20}}{6} = \dfrac{2 \pm 2\sqrt{5}i}{6} = \dfrac{2(1 \pm \sqrt{5}i)}{6} = \dfrac{1 \pm \sqrt{5}i}{3}$

33. $x = \dfrac{2 \pm \sqrt{(-2)^2 - 4(2)(5)}}{2(2)} = \dfrac{2 \pm \sqrt{-36}}{4} = \dfrac{2 \pm 6i}{4} = \dfrac{2(1 \pm 3i)}{4} = \dfrac{1 \pm 3i}{2}$

35. $x = \dfrac{\pi \pm \sqrt{(-\pi)^2 - 4(3)(2)}}{2(3)} = \dfrac{\pi \pm \sqrt{\pi^2 - 24}}{6} = \dfrac{\pi \pm i\sqrt{24 - \pi^2}}{6}$

$\simeq .524 \pm \dfrac{3.759}{6} = .524 \pm .627i$

37. $\sqrt{6}x^2 + \sqrt{6}x - 2.7 = 0$

$x = \dfrac{-\sqrt{6} \pm \sqrt{6 - 4(\sqrt{6})(-2.7)}}{2\sqrt{6}} = -\dfrac{1}{2} \pm \dfrac{\sqrt{32.454}}{2\sqrt{6}} = -.5 \pm 1.163$

39. $x = \dfrac{.2 \pm \sqrt{(.2)^2 - 4(1.1)(.8)}}{2(1.1)} = \dfrac{.2 \pm \sqrt{-3.48}}{2.2} = .091 \pm .848i$

41. $x = \dfrac{3 \pm \sqrt{3^2 - 4(k)1}}{2k} = \dfrac{3 \pm \sqrt{9 - 4k}}{2k}$

$$\left(\dfrac{3 + \sqrt{9-4k}}{2k}\right) + \left(\dfrac{3 - \sqrt{9-4k}}{2k}\right) = \pi$$

$$3 + \sqrt{9-4k} + 3 - \sqrt{9-4k} = 2k\pi$$

$$6 = 2k\pi$$

$$\dfrac{3}{\pi} = k$$

43. $b^2 - 4ac = 1 - 4k8 \geq 0$

$$1 \geq 32k$$

$$\dfrac{1}{32} \geq k$$

45. $x = \dfrac{k \pm \sqrt{k^2 + 20}}{2}$

$$k + 20 = \text{an integer}$$

$$k = \pm 4$$

47. $x = \dfrac{k \pm \sqrt{k^2 - 4(k-1)}}{2} = \dfrac{k \pm \sqrt{k^2 - 4k + 4}}{2}$

$$\dfrac{k + \sqrt{k^2 - 4k + 4}}{2} = \dfrac{k - \sqrt{k^2 - 4k + 4}}{2}$$

$$2\sqrt{k^2 - 4k + 4} = 0$$

$$\sqrt{(k-2)^2} = 0$$

$$k = 2$$

49. $x = \dfrac{2k \pm \sqrt{(-2k)^2 - 4k(5)}}{2k} = \dfrac{2k \pm 2\sqrt{k^2 - 5k}}{2k} = \dfrac{2\left(k \pm \sqrt{k^2 - 5k}\right)}{2k} = \dfrac{k \pm \sqrt{k^2 - 5k}}{k}$

$$\dfrac{k + \sqrt{k^2 - 5k}}{k} = \dfrac{k + \sqrt{k^2 - 5k}}{k} + 3$$

$$2\sqrt{k^2 - 5k} = 3k$$

$$4(k^2 - 5k) = 9k^2$$

$$5k^2 + 20k = 0$$

$$5k(k+4) = 0$$

$$k = \cancel{0}, \ -4$$

51. $x = \dfrac{-b \pm \sqrt{b^2 - 4ac}}{2a}$

$\left(\dfrac{-b + \sqrt{b^2 - 4ac}}{2a} \right)\left(\dfrac{-b - \sqrt{b^2 - 4ac}}{2a} \right) = \dfrac{b^2 - \left(\sqrt{b^2 - 4ac} \right)^2}{4a} = \dfrac{b^2 - (b^2 - 4ac)}{4a} = \dfrac{4ac}{4a^2} = \dfrac{c}{a}$

Section 2.4 Applied Problems That Lead to Quadratic Equations

1. $d^2 + 6d + 7 = 0$

$d = \dfrac{-6 \pm \sqrt{6^2 - 4(1)7}}{2} = \dfrac{-6 \pm \sqrt{8}}{2} = \dfrac{2(-3 \pm \sqrt{2})}{2} = -3 \pm \sqrt{2}$

3. $60 = 20x - x^2$

$x^2 - 20x + 60 = 0$

$x = \dfrac{20 \pm \sqrt{20^2 - 4(1)(60)}}{2} = \dfrac{20 \pm 4\sqrt{10}}{2} = 10 \pm 2\sqrt{10}$

5. $3 = \dfrac{x^2}{5(1 - x)}$

$15(1 - x) = x^2$

$x^2 + 15x - 15 = 0$

$x = \dfrac{-15 \pm \sqrt{15^2 - 4(1)(-15)}}{2} = \dfrac{-15 \pm \sqrt{285}}{2}$

7. $1100 = 100T - T^2$

$T^2 - 100T + 1100 = 0$

$T = \dfrac{100 \pm \sqrt{100^2 - 4(1)(1100)}}{2} = \dfrac{100 \pm \sqrt{5600}}{2} = \dfrac{100 \pm 10\sqrt{56}}{2} = 87.4°, \ 12.6°$

9. $x(5 - x) + 7x = 10$

$5x - x^2 + 7x = 10$

$x^2 - 12x + 10 = 0$

$x = \dfrac{12 \pm \sqrt{12^2 - 40}}{2} = \dfrac{12 \pm \sqrt{104}}{2} = 11.1, \ .9$

11.
$$x^2+(x+7)^2 = 13^2$$
$$x^2+x^2+14x+49 = 169$$
$$2x^2+14x-120 = 0$$
$$x^2+7x-60 = 0$$
$$(x+12)(x-5) = 0$$
$$x=5,\ \cancel{-12}$$
sides: $x=5,\ x+7=12$

13.
$$w=x$$
$$l=x+50$$
$$x(x+50) = 1875$$
$$x^2+50x-1875 = 0$$
$$(x+75)(x-25) = 0$$
$$x=\cancel{-75},\ 25$$
$$w=25,\ l=75$$

15.   Area mowed $=2(x-14)7+2\left(\frac{3}{4}x\right)7$

Area of interior $=\left(\frac{3}{4}x-14\right)(x-14)$

$$14(x-14)+\frac{21}{2}x = \left(\frac{3}{4}x-14\right)(x-14)$$
$$56(x-14)+42x = (3x-56)(x-14)$$
$$56x-784+42x = 3x^2-98x+784$$
$$3x^2-196x+1568 = 0$$
$$x=\frac{196\pm\sqrt{196x^2-4(3)(1568)}}{6} = 56,\ \cancel{9.3}$$

$$l=56,\ w=42$$

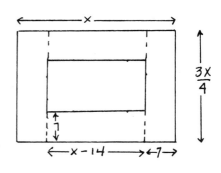

17.

	rate	distance	time
up	$6-x$	20	$20/(6-x)$
down	$6+x$	20	$20/(6+x)$

$$\frac{20}{6-x}+\frac{20}{6+x} = 7.5$$
$$20(6+x)+20(6-x) = 7.5(6-x)(6+x)$$
$$120+20x+120-20x = 270-7.5x^2$$
$$7.5x^2-30 = 0$$
$$x^2-4 = 0$$
$$x = \pm2$$

rate of stream $=2$

19. hours in old week $= x$

   hours in new week $= x-5$

   rate in old week $= \frac{240}{x}$

   rate in new week $= \frac{245}{x-5}$

   $$\frac{240}{x}+1 = \frac{245}{x-5}$$

   $$240(x-5)+x(x-5) = 245x$$

   $$240x-1200+x^2-5x = 245x$$

   $$x^2-10x-1200 = 0$$

   $$(x-40)(x+30) = 0$$

   $$x = \cancel{-30}, 40$$

21. numbers: $x,\ x+1,\ x+2$

   $$x^2+(x+1)^2+(x+2)^2 = 434$$

   $$x^2+x^2+2x+1+x^2+4x+4 = 434$$

   $$3x^2+6x-429 = 0$$

   $$x^2+2x-143 = 0$$

   $$(x+13)(x-11) = 0$$

   $$x = -13,\ 11$$

   numbers: $-13,\ -12,\ -11$ or $11,\ 12,\ 13$

23.
   $$b = 6+h$$

   $$A = \tfrac{1}{2}bh$$

   $$20 = \tfrac{1}{2}(6+h)h$$

   $$h^2+6h-40 = 0$$

   $$(h-4)(h+10) = 0$$

   $$h = 4,\ \cancel{-10}$$

   $$h = 4,\ b = 6+4 = 10$$

25. large pipe $= x$

   small pipe $= x+4$

   $$\frac{1}{x}+\frac{1}{x+4} = \frac{1}{5}$$

   $$5(x+4)+5x = x(x+4)$$

   $$5x+20+5x = x^2+4x$$

   $$x^2-6x-20 = 0$$

   $$x = \frac{6\pm\sqrt{(-6)^2+4(1)(-20)}}{2} = 8.4,\ \cancel{-2.4}$$

27.   units $= x$

    tens $= x-5$

$$2\left(10(x-5)+x\right) = 1+\left(x^2+(x-5)^2\right)$$
$$22x-100 = 1+x^2+x^2-10x+25$$
$$2x^2-32x+126 = 0$$
$$x^2-16x+63 = 0$$

$$x = \frac{16\pm\sqrt{16^2-4(1)(63)}}{2} = \frac{16\pm2}{2} = 7,\, 9$$

    numbers $= 27$ or   49

Section 2.5   Equations in Quadratic Form

1. $(x^2-4)(x^2+3) = 0$

  $(x-2)(x+2)(x^2+3) = 0$

    $x = 2,\, -2$

3. $(x^2-8)(x^2+2) = 0$

  $(x-\sqrt{8})(x+\sqrt{8})(x^2+2) = 0$

    $x = 2\sqrt{2},\, -2\sqrt{2}$

5. $(x^2-8)(x^2+1) = 0$

   $x^2+8 = 0,\ x^2+1 = 0$

   No real solutions

7. $2x^{-2}-5x^{-1}-3 = 0$

   Let $y = x^{-1}$

    $2y^2-5y-3 = 0$

    $(2y+1)(y-3) = 0$

    $y = -\frac{1}{2},\, 3$

    $x = -2,\, \frac{1}{3}$

9. $3x^{-2}-4x^{-1}-4 = 0$

   Let $y = x^{-1}$

    $3y^2-4y-4 = 0$

    $(3y+2)(y-2) = 0$

    $y = -\frac{2}{3},\, 2$

    $x = -\frac{3}{2},\, \frac{1}{2}$

11. $(x+2)^2-2(x+2)-8 = 0$

   Let $y = x+2$

    $y^2-2y-8 = 0$

    $(y-4)(y+2) = 0$

    $y = 4,\, -2$

  $x+2 = 4,$ $x+2 = -2$

    $x = 2$ $x = -4$

13. $(x^2+2)+12(x^2+2)+11 = 0$

   Let $y = x^2+2$

    $y^2+12y+11 = 0$

    $(y+11)(y+1) = 0$

        $y = -11,\, -1$

  $x^2+2 = -11,$ $x^2+2 = -1$

    $x^2 = -13$ $x^2 = -3$

   No real solutions

15. $(x-2)^{-2}+35=12(x-2)^{-1}$

Let $y=(x-2)^{-1}$

$$y^2-12y+35=0$$
$$(y-7)(y-5)=0$$

$$y=7,\ 5$$
$$\frac{1}{x-2}=7,\qquad \frac{1}{x-2}=5$$

$$7x-14=1 \qquad 5x-10=1$$
$$7x=15 \qquad 5x=11$$
$$x=\frac{15}{7} \qquad x=\frac{11}{5}$$

17. $4(x^2+1)^2-7(x^2+1)-2=0$

Let $y=x^2+1$

$$4y^2-7y-2=0$$
$$(4y+1)(y-2)=0$$

$$y=-\frac{1}{4},\ 2$$

$$x^2+1=-\frac{1}{4} \qquad\qquad x^2+1=2$$
$$x^2=-\frac{5}{4} \qquad\qquad x^2-1=0$$

No real solution $\qquad x=-1,\ 1$

19. $\frac{3}{x-1}-\frac{2}{x+3}=\frac{1}{2}$

$$3(2)(x+3)-2(2)(x-1)=(x-1)(x+3)$$
$$6x+18-4x+4=x^2+2x-3$$
$$x^2-25=0$$
$$(x-5)(x+5)=0$$
$$x=+5,\ -5$$

21. $\frac{1}{2}-\frac{1}{x-1}=\frac{6}{(x-1)(x+1)}$

$$(x-1)(x+1)-2(x+1)=6(2)$$
$$x^2-1-2x-2=12$$
$$x^2-2x-15=0$$
$$(x-5)(x+3)=0$$
$$x=-3,\ 5$$

23. $\frac{2}{x-1}+\frac{5}{x-7}=\frac{4}{3x-1}$

$$2(x-7)(3x-1)+5(x-1)(3x-1)=4(x-1)(x-7)$$
$$6x^2-44x+14+15x^2-20x+5=4x^2-32x+28$$
$$17x^2-32x-9=0$$

$$x=\frac{32\pm\sqrt{32^2-4(17)(-9)}}{2(17)}=\frac{32\pm\sqrt{1636}}{34}$$

25. $\dfrac{x}{(x-2)(x+2)}+\dfrac{1}{x-2}=\dfrac{x+5}{(x-2)(x+1)}$

$$x(x-1)+(x+2)(x-1)=(x+5)(x+2)$$
$$x^2-x+x^2+x-2=x^2+7x+10$$
$$x^2-7x-12=0$$
$$x=\frac{7\pm\sqrt{(-7)^2-4(1)(-12)}}{2}=\frac{7\pm\sqrt{97}}{2}$$

27. $\dfrac{x^2-7}{x^2+4}+\dfrac{3}{x^2-1}=\dfrac{x^2+3}{x^2}$

$$(x^2-7)x^2(x^2-1)+3(x^2+4)x^2=(x^2+3)(x^2+4)(x^2-1)$$
$$x^6-7x^4-x^4+7x^2+3x^4+12x^2=x^6+7x^4+12x^2-x^4-7x^2-12$$
$$11x^4-14x^2-12=0$$

Let $y=x^2$

$$11y^2-14y-12=0$$
$$y=\frac{14\pm\sqrt{(-14)^2-4(11)(-12)}}{22}=\frac{14\pm\sqrt{724}}{22}=1.859,\ \ \cancel{-.0587}$$

$$x^2=1.859,\qquad x=\pm1.3636$$

29. $5\sqrt{x-6}=x$

$$25(x-6)=x^2$$
$$x^2-25x+150=0$$
$$(x-10)(x-15)=0$$
$$x=10,\ 15$$
$$5\sqrt{10\ 6}=10\ \checkmark$$
$$5\sqrt{15-6}=15\ \checkmark$$

31. $\sqrt{2y+3}=y$

$$2y+3=y^2$$
$$y^2-2y-3=0$$
$$(y-3)(y+1)=0$$
$$y=\cancel{-1},\ 3$$
$$\sqrt{-2+3}=-1,\ \text{No}$$
$$\sqrt{6+3}=3,\ \checkmark$$

33. $\sqrt{x^2-16}=3$

$$x^2-16=9$$
$$x^2-25=0$$
$$x=5,\ -5$$

35. $\sqrt{z^2+5z+2}=4$

$$z^2+5z+2=16$$
$$z^2+5z-14=0$$
$$(z+7)(z-2)=0$$
$$z=2,\ -7$$
$$\sqrt{4+10+2}=4\ \checkmark$$
$$\sqrt{49-35+2}=4\ \checkmark$$

37.
$$y+1 = \sqrt{3y+7}$$
$$y^2+2y+1 = 3y+7$$
$$y^2-y-6 = 0$$
$$(y-3)(y+2) = 0$$

$$y = 3, \cancel{-2}$$
$$3+1 = \sqrt{9+7} \;\checkmark$$
$$-2+1 = \sqrt{-6+1}, \text{ No}$$

39.
$$2x-1 = \sqrt{2x+5}$$
$$4x^2-4x+1 = 2x+5$$
$$4x^2-6x-4 = 0$$
$$2x^2-3x-2 = 0$$
$$(2x+1)(x-2) = 0$$

$$x = \cancel{-\tfrac{1}{2}}, 2$$

$$-1-1 = \sqrt{-1+5}, \text{ No}$$
$$4-1 = \sqrt{4+5} \;\checkmark$$

41.
$$\sqrt{x+5} = \sqrt{x}-1$$
$$x+5 = x-2\sqrt{x}+1$$
$$4 = -2\sqrt{x}$$
$$16 = 4x$$
$$x = 4$$

$$\sqrt{4+5} = \sqrt{4}-1 \qquad \text{No solution}$$

43.
$$\sqrt{3y-5}-\sqrt{y+7} = 2$$
$$\sqrt{3y-5} = 2+\sqrt{y+7}$$
$$3y-5 = 4+4\sqrt{y+7}+y+7$$
$$y-8 = 2\sqrt{y+7}$$
$$y^2-16y+64 = 4(y+7)$$
$$y^2-20y+36 = 0$$
$$(y-18)(y-2) = 0$$
$$y = \cancel{2}, 18$$

$$\sqrt{6-5}-\sqrt{2+7} = 2, \text{ No}$$
$$\sqrt{54-5}-\sqrt{18+7} = 2 \;\checkmark$$

45.
$$\sqrt{5z+4} = \sqrt{z}+2$$
$$5z+4 = z+4\sqrt{z}+4$$
$$4z = 4\sqrt{z}$$
$$z^2 = z$$
$$z(z-1) = 0$$
$$z = 0, 1$$
$$\sqrt{0+4} = \sqrt{0}+2 \;\checkmark$$
$$\sqrt{5+4} = \sqrt{1}+2 \;\checkmark$$

47.

$$b^2+12^2 = c^2$$
$$b = \sqrt{c^2-144}$$
$$12^2+(b+4)^2 = (c+2)^2$$
$$144+b^2+8b+16 = c^2+4c+4$$
$$144+(c^2-144)+8\sqrt{c^2-144}+16 = c^2+4c+4$$
$$8\sqrt{c^2-144} = 4c-12$$
$$2\sqrt{c^2-144} = c-3$$
$$4(c^2-144) = c^2-6c+9$$
$$3c^2+6c-585 = 0$$
$$c^2+2c-195 = 0$$
$$(c+15)(c-13) = 0$$

$c = \cancel{-15},\ 13$

$a = 12, \quad b = 5, \quad c = 13$

$p = 12+13+5 = 30$

49.

$$S = \pi r\sqrt{r^2+h^2}$$
$$S^2 = \pi^2 r^2(r^2+h^2)$$
$$S^2-\pi^2 r^4 = \pi^2 r^2 h^2$$
$$\frac{S^2-\pi^2 r^4}{\pi^2 r^2} = h^2$$
$$\frac{1}{\pi r}\sqrt{S^2-\pi^2 r^4} = h$$

51.

$$5 = 3+\sqrt{t+2}$$
$$2 = \sqrt{t+2}$$
$$4 = t+2$$
$$t = 2$$
$$5 = 3+\sqrt{4} \ \checkmark$$

Section 2.6   Linear Inequalities; Intervals

17.
$$3x+1 > x-4$$
$$2x > -5$$
$$x > -\frac{5}{2}$$

19.
$$x+1 \geq \sqrt{2}-x\sqrt{3}$$
$$(1+\sqrt{3})x \geq \sqrt{2}-1$$
$$x \geq \frac{\sqrt{2}-1}{\sqrt{3}+1}$$

21.
$$\frac{2x}{3}+5 < \frac{7x}{5}+\frac{1}{3}$$
$$10x+75 < 21x+5$$
$$70 < 11x$$
$$\frac{70}{11} < x$$

23.
$$(x-3)(x+4) > (x-5)(x+5)$$
$$x^2+x-12 > x^2-25$$
$$x-12 > -25$$
$$x > -13$$

– 27 –

25. $x^2-1 \geq (x+3)^2$

$x^2-1 \geq x^2+6x+9$

$-10 \geq 6x$

$-\frac{5}{3} \geq x$

27. $x(x-1) > x(x-3)$

$x^2-x > x^2-3x$

$2x > 0$

$x > 0$

29. $x^2-5x+4 < x(x-2)$

$x^2-5x+4 < x^2-2x$

$-3x < -4$

$x < \frac{4}{3}$

31. $x+2<2x+5\leq6x+7$

$x+2<2x+5$ and $2x+5\leq6x+7$

$-3<x$ $\qquad\qquad$ $-2\leq4x$

$\qquad\qquad\qquad$ $-\frac{1}{2}\leq x$

$-\frac{1}{2}\leq x$

33. $x<2x+3<3x-5$

$x<2x+3$ and $2x+3 < 3x-5$

$-3<x$ $\qquad\qquad$ $8<x$

$x>8$

35. $x^2+1<(x-3)(x-4)\leq(x+3)(x+5)$

$x^2+1<x^2-7x+12\leq x^2+8x+15$

$1 < -7x+12$ and $-7x+12 \leq 8x+15$

$7x < 11$ $\qquad\qquad$ $-15x \leq 3$

$x < \frac{11}{7}$ $\qquad\qquad$ $x \geq -\frac{1}{5}$

$-\frac{1}{5}\leq x<\frac{11}{7}$

41. $-3<x-2<3$

$-3<x-2$ and $x-2 < 3$

$-1<x$ $\qquad\qquad$ $x < 5$

$-1<x<5$

43. $-5<x+2<5$

$-5<x+2$ and $x+2 < 5$

$-7<x$ $\qquad\qquad$ $x < 3$

$-7<x<3$

45. $x-5 >1$ or $x-5 < -1$

$x > 6$ $\qquad\qquad$ $x < 4$

$(6, \infty) \cup (-\infty, 4)$

47. $-5 \leq \frac{1}{2}x - 3 \leq 5$

$\quad -10 \leq x - 6 \quad$ and $\quad x - 6 \leq 10$

$\quad -4 \leq x \qquad\qquad\qquad x \leq 16$

49. $-2 \leq 2x - 3 \leq 2$

$\quad -2 \leq 2x - 3 \quad$ and $\quad 2x - 3 \leq 2$

$\quad 1 \leq 2x \qquad\qquad\qquad 2x \leq 5$

$\quad \frac{1}{2} \leq x \qquad\qquad\qquad x \leq \frac{5}{2}$

55. $2 < x < 6$

$\quad m = \frac{6+2}{2} = 4$

$\quad hl = \frac{|2-6|}{2} = 2$

$\quad |x-4| < 2$

57. $-10 < x < 16$

$\quad m = \frac{-10+16}{2} = 13$

$\quad hl = \frac{|-10-16|}{2} = 13$

$\quad |x-3| < 13$

61. $2 \leq x \leq 7$

$\quad m = \frac{2+7}{2} = \frac{9}{2}$

$\quad hl = \frac{|2-7|}{2} = \frac{5}{2}$

$\quad \left| x - \frac{9}{2} \right| \leq \frac{5}{2}$

$\quad |2x-9| \leq 5$

65. $|M-10| < .01$

$\quad -.01 < M - 10 < .01$

$\quad 9.99 < M < 10.01$

1. $x^2+2x \geq 3$

$x^2+2x-3 \geq 0$

$(x+3)(x-1) \geq 0$

| $x+3$ | $-$ | $+$ | $+$ |
| $x-2$ | $-$ | $-$ | $+$ |

$-3 \qquad 1$

$x \leq -3 \quad$ or $\quad x \geq 1$

3. $5x < 3x^2-2$

$3x^2-5x-2 > 0$

$(3x+1)(x-1) > 0$

| $3x+1$ | $-$ | $+$ | $+$ |
| $x-2$ | $-$ | $-$ | $+$ |

$-1/3 \qquad 2$

$x < -\frac{1}{3} \quad$ or $\quad x > 2$

5. $x(6-x) \geq 3x^2+2x-15$

$0 \geq 4x^2-4x-15$

$0 \geq (2x+3)(2x-5)$

| $2x+3$ | $-$ | $+$ | $+$ |
| $2x-5$ | $-$ | $-$ | $+$ |

$-\frac{3}{2} \qquad \frac{5}{2}$

$-\frac{3}{2} \leq x \leq \frac{5}{2}$

7. $x-1 < x^2+1$

$0 < x^2-x+2$

$x < \dfrac{1 \pm \sqrt{1-8}}{2}$

All x

9. $x+4 < x^2-4$

$0 < x^2-x-8$

$0 < \left(x - \dfrac{\left(1+\sqrt{33}\right)}{2}\right)\left(x + \dfrac{\left(1-\sqrt{33}\right)}{2}\right)$

| $x - \dfrac{(1+\sqrt{33})}{2}$ | $-$ | $-$ | $+$ |
| $x - \dfrac{(1-\sqrt{33})}{2}$ | $-$ | $+$ | $+$ |

$\dfrac{1-\sqrt{33}}{2} \qquad \dfrac{1+\sqrt{33}}{2}$

$x < \dfrac{1-\sqrt{33}}{2} \quad$ or $\quad x > \dfrac{1+\sqrt{33}}{2}$

11. $\dfrac{(x+4)(x+5)}{x+1}$

$x+4$	$-$	$-$	$+$	$+$
$x+5$	$-$	$+$	$+$	$+$
$x+1$	$-$	$-$	$-$	$+$

$$\begin{array}{ccccccc} \hline & & | & & | & & | \\ & & -5 & & -4 & & -1 \end{array}$$

$x<-5$ or $-4<x<-1$

13. $\dfrac{x+1}{2x+3}-1 \le 0$

$$\dfrac{x+1-(2x+3)}{2x+3} \le 0$$

$$\dfrac{-(x+2)}{2x+3} \le 0$$

$-(x+2)$	$+$	$-$	$-$
$2x+3$	$-$	$-$	$+$

$$\begin{array}{ccccc} \hline & & | & & | \\ & & -2 & & -\frac{3}{2} \end{array}$$

$x\le-2$ or $x>-\frac{3}{2}$

15. $\dfrac{(x-1)(x+1)(x^2+1)}{(x-2)(x+2)}\le 0$

$x-1$	$-$	$-$	$-$	$+$	$+$
$x+1$	$-$	$-$	$+$	$+$	$+$
x^2+1	$+$	$+$	$+$	$+$	$+$
$x-2$	$-$	$-$	$-$	$-$	$+$
$x+2$	$-$	$+$	$+$	$+$	$+$

$$\begin{array}{ccccccccc} \hline & & | & & | & & | & & | \\ & & -2 & & -1 & & 1 & & 2 \end{array}$$

$-2<x\le-1$ or $1\le x<2$

17. $\dfrac{x^2+1}{x^2+4}-1 \le 0$

$$\dfrac{x^2+1-(x^2+4)}{x^2+4} \le 0$$

$$\dfrac{-3}{x^2+4} \le 0$$

All x

19. $\dfrac{x-2}{x^2+x-1}\ge 0$

$$\dfrac{x-2}{\left(x-\dfrac{(-1+\sqrt{5})}{2}\right)\left(x-\dfrac{(-1-\sqrt{5})}{2}\right)}\ge 0$$

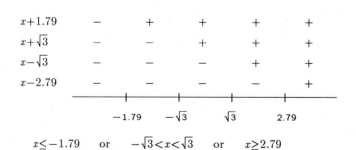

$x-2$	$-$	$-$	$-$	$+$
$x-\dfrac{\left(-1+\sqrt{5}\right)}{2}$	$-$	$-$	$+$	$+$
$x-\dfrac{\left(-1-\sqrt{5}\right)}{2}$	$-$	$+$	$+$	$+$

$$\dfrac{-1-\sqrt{5}}{2}<x<\dfrac{-1+\sqrt{5}}{2} \quad\text{or}\quad x\ge 2$$

21. $\dfrac{x+2}{x^2-3}\le 1$

$$\dfrac{x+2-(x^2-3)}{x^2-3}\le 0$$

$$\dfrac{-x^2+x+5}{x^2-3}\le 0$$

$$\dfrac{x^2-x-5}{x^2-3}\ge 0$$

$$\dfrac{(x+1.79)(x-2.79)}{(x-\sqrt{3})(x+\sqrt{3})}\ge 0$$

$x+1.79$	$-$	$+$	$+$	$+$	$+$
$x+\sqrt{3}$	$-$	$-$	$+$	$+$	$+$
$x-\sqrt{3}$	$-$	$-$	$-$	$+$	$+$
$x-2.79$	$-$	$-$	$-$	$-$	$+$

$$x\le -1.79 \quad\text{or}\quad -\sqrt{3}<x<\sqrt{3} \quad\text{or}\quad x\ge 2.79$$

23. $4x^2 - kx + 3 = 0$

$$k^2 - 4(4)(3) \geq 0$$
$$k^2 - 48 \geq 0$$
$$(k - \sqrt{48})(k + \sqrt{48}) \geq 0$$

$k - \sqrt{48}$	$-$	$-$	$+$
$k + \sqrt{48}$	$-$	$+$	$+$

$-\sqrt{48}$ $\sqrt{48}$

$$k \leq -\sqrt{48} \quad \text{or} \quad k \geq \sqrt{48}$$

25. $15t - 3t^2 > 0$

$$3t(5 - t) > 0$$

$3t$	$-$	$+$	$+$
$5 - t$	$+$	$+$	$-$

0   5

$$0 < t < 5$$

27. $500 + 600T - 20T^2 > 4500$

$$20T^2 - 600T + 4500 < 0$$
$$T^2 - 30T + 200 < 0$$
$$(T - 10)(T - 20) < 0$$

$T - 10$	$-$	$+$	$+$
$T - 20$	$-$	$-$	$+$

10   20

$$10 < T < 20$$

Chapter 2   Review

1. $2x - 3(x - 5) = 12$

$$2x - 3x + 15 = 12$$
$$x = 3$$

3. $\frac{3x}{2} + 2(3x - 5) = \frac{x}{3}$

$$9x + 12(3x - 5) = 2x$$
$$43x = 60$$
$$x = \frac{60}{43}$$

5. $x - 3a(6 - 7x) = 19ax$

$$x - 18a + 21ax = 19ax$$
$$2ax + x = 18a$$
$$x(2a + 1) = 18a$$
$$x = \frac{18a}{2a + 1}$$

7. $\frac{2}{x} + \frac{3}{2x} = 5$

$$4 + 3 = 10x$$
$$\frac{7}{10} = x$$

9. $\dfrac{2}{2y-3}=\dfrac{5}{y+4}$

       $2(y+4)=5(2y-3)$

            $23=8y$

            $\dfrac{23}{8}=y$

11. $\dfrac{4}{t-5}+3-\dfrac{3t}{t+4}=0$

      $4(t+4)+3(t+4)(t-5)-3t(t-5)=0$

      $4t+16+3t-3t-60-3t^2+15t=0$

                $16t=44$

                  $t=\dfrac{11}{4}$

13. $|x|+3=7$

        $|x|=4$

         $x=\pm4$

15. $|x-4|=10$

     $x\ge4$ $x<4$

    $x-4=10$ $-(x-4)=10$

      $x=14$ $x=-6$

17. $2y+|y-7|=5$

    $y-7>0,\ y>7$ $y-7<0,\ y<7$

    $2y+y-7=5$ $2y-(y-7)=5$

        $3y=12$ $y+7=5$

         $y=4$ $y=-2$

       Impossible

19. $3x^2-7x=0$

      $x(3x-7)=0$

        $x=0,\ \dfrac{7}{3}$

21. $(y-\sqrt{10})(y+\sqrt{10})=0$

            $y=\pm10$

23. $7a=2a^2+3$

      $2a^2-7a+3=0$

    $(2a-1)(a-3)=0$

        $a=\dfrac{1}{2},\ 3$

25. $z^2+6z+5=0$

    $(z+5)(z+1)=0$

       $z=-5,\ -1$

27. $y=\dfrac{4\pm\sqrt{(-4)^2-4(4)(-11)}}{2(4)}=\dfrac{4\pm8\sqrt{3}}{8}=\dfrac{1\pm2\sqrt{3}}{2}$

29. $x=\dfrac{12\pm\sqrt{(-12)^2-4(2)(-15)}}{4}=\dfrac{12\pm\sqrt{264}}{4}=\dfrac{6\pm\sqrt{66}}{2}$

31. $15y^2+14y-8=0$

$$y=\frac{-14\pm\sqrt{14^2-4(15)(-8)}}{2(15)}=\frac{-14\pm\sqrt{676}}{30}=-\frac{4}{3},\frac{2}{5}$$

33. $$t=\frac{4\pm\sqrt{(-4)^2-4(1)(8)}}{2}=\frac{4\pm\sqrt{-16}}{2}=2\pm2i$$

35. $$x=\frac{-7\pm\sqrt{(7)^2-4(3)(-2)}}{2(3)}=\frac{-7\pm\sqrt{73}}{6}$$

37. $h^4-7h^2+12=0$

Let $y=h^2$

$y^2-7y+12=0$

$(y-3)(y-4)=0$

$y=3,\,4$

$h^2=3,\,4$

$h=\pm\sqrt{3},\,\pm2$

39. $3(x-1)^2-10(x-1)-8=0$

$y=x-1$

$3y^2-10y-8=0$

$(3y+2)(y-4)=0$

$y=-\frac{2}{3},\,4$

$x-1=-\frac{2}{3},\qquad x-1=4$

$x=\frac{1}{3},\qquad\qquad x=5$

41. $\frac{1}{t+1}+\frac{1}{t}=\frac{9}{20}$

$20t+20(t+1)=9t(t+1)$

$9t^2-31t-20=0$

$(9t+5)(t-4)=0$

$t=-\frac{5}{9},\,4$

43. $\sqrt{x+12}-\sqrt{x}=3$

$\sqrt{x+12}=\sqrt{x}+3$

$x+12=x+6\sqrt{x}+9$

$3=6\sqrt{x}$

$\frac{1}{2}=\sqrt{x}$

$\frac{1}{4}=x$

45. $\sqrt{7t+23} - \sqrt{3t+7} = 2$

$\qquad \sqrt{7t+23} = 2 + \sqrt{3t+7}$

$\qquad 7t+23 = 4 + 4\sqrt{3t+7} + 3t+7$

$\qquad 4t+12 = 4\sqrt{3t+7}$

$\qquad t+3 = \sqrt{3t+7}$

$\qquad t^2 + 6t + 9 = 3t + 7$

$\qquad t^2 + 3t + 2 = 0$

$\qquad (t+2)(t+1) = 0$

$\qquad t = -1, \ -2$

$\qquad \sqrt{-7+23} - \sqrt{-3+4} = 2 \quad \checkmark$

$\qquad \sqrt{-14+23} - \sqrt{-6+7} = 2 \quad \checkmark$

47. $x - 4 > 2x + 3$

$\qquad -7 > x$

49. $3x + 5 \leq 2 - 3x$

$\qquad 6x \leq -3$

$\qquad x \leq -\frac{1}{2}$

51. $x(x-5) < (x+1)(x-2)$

$\qquad x^2 - 5x < x^2 - x - 2$

$\qquad 2 < 4x$

$\qquad \frac{1}{2} < x$

53. $3 \leq 2x - 5 < 8$

$\qquad 3 \leq 2x-5 \qquad 2x-5 < 8$

$\qquad 8 \leq 2x \qquad\qquad 2x < 13$

$\qquad 4 \leq x \qquad\qquad x < \frac{13}{2}$

$\qquad\qquad 4 \leq x < \frac{13}{2}$

55. $x - 1 < 2x < 3x + 5$

$\quad x-1 < 2x \quad$ and $\quad 2x < 3+5$

$\quad -1 < x \qquad\qquad -5 < x$

$\qquad x > -1$

57. $x + 3 \geq 0, \ x \geq -3 \quad$ or $\quad x+3 < 0, \ x < -3$

$\quad x + 3 > 1 \qquad\qquad\quad -(x+3) > 1$

$\qquad x > -2 \qquad\qquad\qquad x + 3 < -1$

$\qquad\qquad\qquad\qquad\qquad\quad x < -4$

 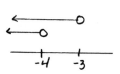

$\qquad\qquad\qquad\qquad x > -2 \quad$ or $\quad x < -4$

59. $-2 < x < 5$

$$m = \frac{-2+5}{2} = \frac{3}{2}$$

$$hl = \frac{|-2-5|}{2} = \frac{7}{2}$$

$$\left|x - \frac{3}{2}\right| < \frac{7}{2}$$

$$|2x-3| < 7$$

61. $x \le -1 \quad$ or $\quad x \ge 3$

$$m = \frac{-1+3}{2} = 1$$

$$hl = \frac{|-1-3|}{2} = 2$$

$$|x-1| \ge 2$$

63. $\quad x^2 + 5x - 6 \le 0$

$$(x+6)(x-1) \le 0$$

$x+6$	$-$	$+$	$+$
$x-1$	$-$	$-$	$+$

$$-61$$

$$-6 \le x \le 1$$

65. $\quad x+2 > x^2 + 2x$

$$0 > x^2 + x - 2$$

$$0 > (x+2)(x-1)$$

$x+2$	$-$	$+$	$+$
$x-1$	$-$	$-$	$+$

$$-21$$

$$-2 < x < 1$$

67. $\quad \dfrac{x-5}{(x-3)(x-5)} \ge 0$

$$\frac{1}{x-3} \ge 0$$

$x-3$	$-$	$+$

$$3$$

$$x > 3, \ x \ne 5$$

69. $\quad P_1 - P_2 = \frac{1}{2}\rho\left(v_2{}^2 - v_1{}^2\right)$

$$\frac{2(P_1 - P_2)}{v_2{}^2 - v_1{}^2} = \rho$$

71. $\quad E = c^2(m - m_0)$

$$E = c^2 m - c^2 m_0$$

$$c^2 m_0 = c^2 m - E$$

$$m_0 = \frac{c^2 m - E}{c^2}$$

73. $x + 3x = 787$

$$4x = 787$$

$$x = 196.75$$

$$3x = 590.25$$

75.

	amt.	concentration
S_1	x	.06
S_2	$10-x$	.02
$S_1 + S_2$	10	.05

$$.06x + .02(10-x) = .05(10)$$

$$4x = 30 \qquad x = 7.5 \text{ of } 6\% \qquad 10 - x = 2.5 \text{ of } 2\%$$

77.
$$v = 5t - 16t^2$$
$$-100 = 5t - 16t^2$$
$$16t^2 - 5t - 100 = 0$$
$$t = \frac{5 \pm \sqrt{25 - 4(16)(-100)}}{32}$$
$$t = 2.66, \; -2.35$$

79.
$$\frac{1}{p} + \frac{1}{p-3} = \frac{1}{3}$$
$$3(p-3) + 3p = p(p-3)$$
$$p^2 - 9p + 9 = 0$$
$$p = \frac{9 \pm \sqrt{9^2 - 4(1)(9)}}{2}$$
$$= \frac{9 \pm 3\sqrt{5}}{2}$$

81.
$$m = m_0\left(1 - \frac{v^2}{c^2}\right)^{-\frac{1}{2}}$$
$$\frac{m}{m_0} = \left(\frac{c^2 - v^2}{c^2}\right)^{-\frac{1}{2}}$$
$$\frac{m}{m_0} = \left(\frac{c^2}{c^2 - v^2}\right)^{\frac{1}{2}}$$
$$\frac{m^2}{m_0{}^2} = \frac{c^2}{c^2 - v^2}$$
$$m^2 c^2 - m^2 v^2 = m_0{}^2 c^2$$
$$-m^2 v^2 = m_0{}^2 c^2 - m^2 c^2$$
$$v^2 = \frac{c^2(m_0{}^2 - m^2)}{-m^2}$$
$$v^2 = \frac{c^2(m^2 - m_0{}^2)}{m^2}$$
$$v = c\left(1 - \frac{m_0{}^2}{m^2}\right)^{\frac{1}{2}}$$

Chapter 3 The Idea of a Function

33. $d = \sqrt{\left(0-(-1)\right)^2 + (4-3)^2} = \sqrt{2}$

  $p = \left(\frac{0-1}{2}, \frac{4+3}{2}\right) = \left(-\frac{1}{2}, \frac{7}{2}\right)$

35. $d = \sqrt{\left(\frac{1}{2}-\frac{1}{2}\right)^2 + \left(\frac{1}{2}-(-\frac{3}{4})\right)^2} = \frac{5}{4}$

  $p = \left(\frac{\frac{1}{2}+\frac{1}{2}}{2}, \frac{\frac{1}{2}-\frac{3}{4}}{2}\right) = \left(\frac{1}{2}, -\frac{1}{8}\right)$

37. $d = \sqrt{(.5-6.2)^2 + (1.6-7.5)^2} = 8.2$

  $p = \left(\frac{.5+6.2}{2}, \frac{1.6+7.5}{2}\right) = (3.35,\ 4.55)$

39. $d = \sqrt{\left(2-(-\sqrt{3})\right)^2 + \left(-6-(-3)\right)^2} = \sqrt{4 + 4\sqrt{3} + 3 + 9} = \sqrt{16 + 4\sqrt{3}} = 2\sqrt{4 + \sqrt{3}}$

  $p = \left(\frac{2-\sqrt{3}}{2}, \frac{-6-3}{2}\right) = \left(\frac{2-\sqrt{3}}{2}, -\frac{9}{2}\right)$

41. $d = \sqrt{\left(\sqrt{x}-(-\sqrt{x})\right)^2 + \left(\sqrt{y}-(-\sqrt{y})\right)^2} = \sqrt{(2\sqrt{x})^2 + (2\sqrt{y})^2} = \sqrt{4x + 4y} = 2\sqrt{x + y}$

43. $d = \sqrt{(x-x)^2 + \left(y-(-y)\right)^2} = \sqrt{(2y)^2} = 2|y|$

45. $\sqrt{(-4-0)^2 + (-1-y)^2} = \sqrt{(-1-0)^2 + (2-y)^2}$

  $16 + 1 + 2y + y^2 = 1 + 4 - 4y + y^2$

  $6y = -12$

  $y = -2$

Section 3.2   Variation

11. map distance $= x$

real distance $= y$

$$y = 20x$$

$$y = 20(12.6) = 252$$

13.   sodium $= x$

chlorine $= 100 - x$

$$\frac{x}{100-x} = \frac{5}{3}$$

$$3x = 500 - 5x$$

$$8x = 500$$

$$x = 62.5$$

$$100 - x = 37.5$$

15. $x,\ 8-x$

$$\frac{x}{8-x} = \frac{5}{4}$$

$$4x = 40 - 5x$$

$$9x = 40$$

$$x = 4.44$$

$$8 - x = 3.56$$

17.   dealer's cost $= x$

list cost $= l$

$$l = 1.18x$$

$$8266 = 1.18x$$

$$x = \$7005$$

21. $y = kx$

$$5 = k10$$

$$k = \frac{1}{2}$$

$$y = \frac{1}{2}x$$

$$y = \frac{1}{2}(12) = 6$$

23. $z = \frac{k}{t}$

$$4 = \frac{k}{\frac{1}{2}}$$

$$k = 2$$

$$z = \frac{2}{t}$$

$$z = \frac{2}{2} = 1$$

25. $v = kt$

$$96 = k3$$

$$k = 32$$

$$v = 32t$$

$$v = 32(2) = 64$$

27.   cost of production $= x$

circulation $= y$

$$x = \frac{k}{\sqrt{y}}$$

$$.40 = \frac{k}{\sqrt{200,000}}, \quad k = 178.89$$

$$x = \frac{178.89}{\sqrt{y}}$$

$$x = \frac{178.89}{\sqrt{400,000}} = .283$$

29. $I = \dfrac{k}{s^2}$

$4 = \dfrac{k}{(5)^2}$

$k = 100$

$I = \dfrac{100}{s^2}$

$I = \dfrac{100}{100} = 1$

31. $p = kri^2$

$50 = k50(.5)^2$

$k = \dfrac{1}{(.5)^2} = 4$

$p = 4ri^2$

$p = 4(50)1^2 = 200$

33. $I = \dfrac{kV}{R}$

$.5 = \dfrac{k(10)}{20}$

$k = 1$

$I = \dfrac{V}{R}$

$I = \dfrac{12}{20} = .6$

35. $C = 2000 + kn$

$2200 = 2000 + k(1000)$

$k = .2$

$C = 2000 + .2n$

$C = 2000 + .2(1500) = 2300$

37. $y = kt,\ x = ct$

    (a) $xy = (ct)(kt) = ckt^2$

          xy varies directly as t^2

    (b) $\dfrac{x}{y} = \dfrac{ct}{kt} = \dfrac{c}{k}$

          $\dfrac{x}{y}$ is a constant

    (c) $x + y = ct + kt = (c + k)t$

          $x + y$ varies directly as t

Section 3.3   Functions

21. $f(x) = \sqrt{-x}$

      $-x \geq 0,\ x \leq 0$: Domain

            $y \geq 0$: Range

23. $f(x) = \sqrt{x-1}$

      $x - 1 \geq 0,\ x \geq 1$: Domain

            $y \geq 0$: Range

25. (a) $f(3)=3(3)+1=10$

(b) $f(\pi)=3\pi+1$

(c) $f(z)=3z+1$

(d) $f(x-h)=3(x-h)+1$

(e) $10 = 3x+1$

$9 = 3x$

$3 = x$

(f) All reals

(g) $\dfrac{f(a+h)-f(a)}{h}=\dfrac{3(a+h)+1-(3a+1)}{h}=\dfrac{3a+3h+1-3a-1}{h}=\dfrac{3h}{h}=3$

27. $f(x)=x^2$

$$\frac{f(a+h)-f(a)}{h}=\frac{(a+h)^2-a^2}{h}=\frac{a^2+2ah+h^2-a^2}{h}=\frac{h(2a+h)}{h}=2a+h$$

29. $f(x)=x^2-2x$

$$\frac{f(a+h)-f(a)}{h}=\frac{(a+h)^2-2(a+h)-(a^2-2a)}{h}$$

$$=\frac{a^2+2ah+h^2-2a-2h-a^2+2a}{h}=\frac{h(2a+h-2)}{h}=2a+h-2$$

31. $f(x)=x^3-1$

$$\frac{f(a+h)-f(a)}{h}=\frac{(a+h)^3-1-(a^3-1)}{h}=\frac{a^3+3a^2h+3ah^2+h^3-1-a^3+1}{h}$$

$$=\frac{h(3a^2+3ah+h^2)}{h}=3a^2+3ah+h^2$$

33. $f(-x)=-3(-x)^5=3\times5=-f(x)$, odd

35. $f(-x)=2(-x)-(-x)^3=-2x+x^3=-(2x-x^3)=-f(x)$, odd

37. $f(-x)=(-x)^4-3(-x)=x^4-3x$, neither

39. $f(-x)=\sqrt{(-x)^2+4}=\sqrt{x^2+4}=f(x)$, even

41. $g(x) = 2x - 7$

$g(x-4) = 2(x-4) - 7 = 2x - 15$

$g(x-1) = 2(x-1) - 7 = 2x - 9$

$g(x-4) \cdot g(x-1) = (2x-15)(2x-9) = 4x^2 - 48x + 135$

43. $F(x) = 3x^2$

$F(x-3) = 3(x-3)^2 = 3(x^2 - 6x + 9) = 3x^2 - 18x + 27$

$F(x) \cdot F(x-3) = 3x^2(3x^2 - 18x + 27) = 9x^4 - 54x^3 + 81x^2$

45. $f(x) = x^2 + 2x$

$f(x) - 3 = x^2 + 2x - 3 = 0$

$(x+3)(x-1) = 0$

$x = 1, \ -3$

47. $f(x) = 2x - x^2$

$f(x) - f(2x) = 2x - x^2 - \left((2)(2x) - (2x)^2 \right) = 0$

$2x - x^2 - 4x + 4x^2 = 0$

$3x^2 - 2x = 0$

$x(3x-2) = 0$

$x = 0, \frac{2}{3}$

49. $f(x) = \frac{k}{x}$

(a) $\dfrac{f(x_1)}{f(x_2)} = \dfrac{\frac{k}{x_1}}{\frac{k}{x_2}} = \dfrac{k}{x_1} \cdot \dfrac{x_2}{k} = \dfrac{x_2}{x_1}$

(b) $f(\frac{1}{x}) = \dfrac{k}{\frac{1}{x}} = kx$

$\dfrac{1}{f(x)} = \dfrac{1}{\frac{k}{x}} = \dfrac{x}{k}$

(c) $f(x^2) = \dfrac{k}{x^2}$

$\left[f(x) \right]^2 = (\frac{k}{x})^2 = \dfrac{k^2}{x^2}$

(d) $\dfrac{f(a+h) - f(a)}{h} = \dfrac{\frac{k}{a+h} - \frac{k}{a}}{h} = \dfrac{\frac{ka - k(a+h)}{(a+h)a}}{h} = \dfrac{-kh}{(a+h)ah} = \dfrac{-k}{(a+h)a}$

(problem #49, continued)

(e) $f(x)+1=\frac{k}{x}+1=\frac{k+x}{x}$

$$f(x+1)=\frac{k}{x+1}$$

(f) $f(x_1+x_2)=\frac{k}{x_1+x_2}$

$$f(x_1)+f(x_2)=\frac{k}{x_1}+\frac{k}{x_2}=\frac{kx_2+kx_1}{x_1x_2}$$

(g) $af(x)=a(\frac{k}{x})=\frac{ak}{x}$

$$f(ax)=\frac{k}{ax}$$

51. $f(s)=3.101s^2+29.46s$

$f(-.415)=3.101(-.415)^2+29.46(-.415)$

$=-11.6918$

53. $i(t)=1.196t^2+.076t$

$i(1.375)=1.196(1.375)^2+.076(1.375)$

$=2.366$

55. $p=w-.12w-.10w-.06w-.015w-150$

$=.705w-150$

Section 3.4   The Graph of a Function

17. $p=z^2-z-6$

$0=(z-3)(z+2)$

$z=3, -2$, (zeros)

23. $y(x)=\sqrt{x-2}$

$-2\geq0, x\geq2:$ Domain

$y\geq0:$ Range

$0=\sqrt{x-2}$

$x=2$, (zero)

19. $y=\sqrt{16-4x^2}$

$16-4x^2\geq0$

$(4-2x)(4+2x)\geq0$

$4-2x$	$+$	$+$	$-$
$4+2x$	$-$	$+$	$+$

-2 2

Domain: $-2\leq x\leq2$

Range: $4\geq y\geq0$

25. $v = s^3 - 4s^2$

$0 = s^2(s-4)$

$s = 0, 4$ (zeros)

Section 3.5   Composite and Inverse Functions

1. $f \circ g = f(3-2)x = (3-2x)-5 = -2-2x$

$g \circ f = g(x-5) = 3-2(x-5) = 13-2x$

3. $f \circ g = f(2x) = 3(2x)+4 = 6x+4$

$g \circ f = g(3x+4) = 2(3x+4) = 6x+8$

5. $f \circ g = f(x^2) = x^2$

$g \circ f = g(x) = x^2$

7. $f \circ g = f\left(x^{1/3}\right) = \left(x^{1/3}\right)^3 - 1 = x-1$

$g \circ f = g(x^3-1) = (x^3-1)^{1/3}$

9. $2x+3 = 6,$

$2x = 3$

$x = \dfrac{3}{2}$

11. $10+4t = 20$

$4t = 10$

$t - \dfrac{5}{2}$

13.

15.

17.

19.

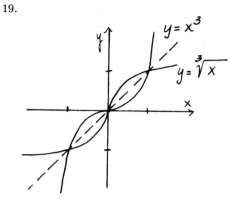

21. $f \circ g = f(\frac{-x}{2}) = -2(\frac{-x}{2}) = x$

23. $f \circ g = f(\frac{-x}{3}+2) = 6-3(\frac{-x}{3}+2) = x$

25. $f \circ g = f(\frac{x}{5}-1) = 5\left((\frac{x}{5}-1)+1\right) = x$

27. $f\Big(g(x)\Big) = x$

$g(x) - 3 = x$

$g(x) = x + 3$

29. $f\Big(g(x)\Big) = x$

$\dfrac{1}{g(x)+1} = x$

$g(x) + 1 = \frac{1}{x}$

$g(x) = \frac{1}{x} - 1$

$= \frac{1-x}{x}$

31. $f\Big(g(x)\Big) = x$

$\dfrac{g(x)-1}{g(x)+1} = x$

$g(x) - 1 = xg(x) + x$

$g(x) - xg(x) = x + 1$

$g(x) = \frac{x+1}{1-x}$

3. $d=\sqrt{\left(3-(-2)\right)^2+\left(7-(-1)\right)^2}=\sqrt{5^2+8^2}=\sqrt{89}$

5. $p=\frac{F}{a}$, $p=\frac{16}{2.4}=6.67$

7. $C=\frac{v}{n}$, $C=\frac{300{,}000}{750{,}000}=.40$

9. $B=\frac{kwd^2}{h}$, $2500=\frac{k(6)(8^2)}{12}$, $k=\frac{2500}{32}=\frac{625}{8}$

$B=\frac{\frac{625}{8}(6)(8^2)}{15}=2000$

23. $g(x)=\sqrt{x-3}$, $x-3\geq0$, $\quad\begin{array}{ll}x\geq3: & \text{Domain}\\ g(x)\geq0: & \text{Range}\end{array}$

25. $y=4-2x$, $4-2x=0$, $x=2$

27. $F(x)=-\sqrt{x}+2$, $0=-\sqrt{x}+2$, $\sqrt{x}=2$, $x=4$

29. $y=\sqrt{-x}$, $0=\sqrt{-x}$, $x=0$

33. $f\left(g(x)\right)=x$

$2g(x)+5=x$

$g(x)=\frac{x-5}{2}$

35. $g\left(h(x)\right)=x$

$\frac{1}{h(x)+2}=x$

$1=xh(x)+2x$

$h(x)=\frac{1}{x}-2=\frac{1-2x}{x}$

37. $f\circ g=f(4-x)=3(4-x)-1=11-3x$

$g\circ f=g(3x-1)=4-(3x-1)=5-3x$

39. $f\circ g=f(\frac{1}{x})=\dfrac{1}{\frac{1}{x}-1}=\dfrac{1}{\frac{1-x}{x}}=\dfrac{x}{1-x}$

$g\circ f=g\left(\frac{1}{x-1}\right)=\dfrac{1}{\frac{1}{x-1}}=x-1$

Chapter 4 Elementary Functions

Section 4.1   Linear Functions

13. $m = \frac{2-4}{1-5} = \frac{-2}{-4} = \frac{1}{2}$

15. $m = \frac{-1-(-6)}{-1-3} = \frac{5}{-4} = -\frac{5}{4}$

17. $m = \frac{-3-(-7)}{-2-(-5)} = \frac{4}{3}$

19. $m = \frac{3-3}{-2-5} = 0$

21. $y-5 = \frac{1}{2}(x-2)$

$2y-10 = x-2$

$-8 = x-2y$

23. $y-(-2) = -7(x-5)$

$y+2 = -7x+35$

$y+7x = 33$

25. $y-3 = \left(\frac{3-2}{1-6}\right)(x-1)$

$y-3 = -\frac{1}{5}(x-1)$

$5y-15 = -x+1$

$5y+x = 16$

27. $y-2 = \left(\frac{-1-2}{-1-1}\right)(x-1)$

$y-2 = \frac{3}{2}(x-1)$

$2y-4 = 3x-3$

$2y-3x = 1$

29. $y-2 = \left(\frac{2-0}{0-(-5)}\right)(x-0)$

$y-2 = \frac{2}{5}x$

$5y-10 = 2x$

$5y-2x = 10$

31. $2x-3y = 5$

$3y = 2x-5$

$y = \frac{2x}{3} - \frac{5}{3}$

$m = \frac{2}{3},\ b = -\frac{5}{3}$

33. $x+y = 2$

$y = -x+2$

$m = -1,\ b = 2$

35. $4y-2x+8 = 0$

$4y = 2x-8$

$y = \frac{1}{2}x-2$

$m = \frac{1}{2},\ b = -2$

37. $2y = x+5$

$\qquad y = \frac{x}{2} + \frac{5}{2}$

$\qquad m = \frac{1}{2}, \ b = \frac{5}{2}$

41. $x - 5y + 7 = 0$

$\qquad 5y = x + 7$

$\qquad y = \frac{x}{5} + \frac{7}{5}$

$\qquad m = \frac{1}{5}, \ b = \frac{7}{5}$

43. $3x + 2y = 7$

$\qquad 2y = -3x + 7$

$\qquad y = -\frac{3}{2}x + \frac{7}{2}$

$\qquad m = -\frac{3}{2}, \ m' = \frac{2}{3}$

$\qquad y - 2 = \frac{2}{3}(x-1)$

$\qquad 3y - 6 = 2x - 2$

$\qquad 3y - 2x = 4$

45. $x - y = 2$

$\qquad y = x - 2$

$\qquad m = 1, \ m' = -1$

$\qquad y - 3 = -1(x-5)$

$\qquad y + x = 8$

47. $2y - 3x = 1$

$\qquad 2y = 3x + 1$

$\qquad y = \frac{3}{2}x + \frac{1}{2}$

$\qquad m = \frac{3}{2}, \ m' = -\frac{2}{3}$

$\qquad y - 2 = -\frac{2}{3}(x-1)$

$\qquad 3y - 6 = -2x + 2$

$\qquad 3y + 2x = 8$

49. $5x - 7y = 0$

$\qquad 7y = 5x$

$\qquad y = \frac{5}{7}x$

$\qquad m = \frac{5}{7}, \ m' = -\frac{7}{5}$

$\qquad y - 0 = -\frac{7}{5}(x-0)$

$\qquad 5y = -7x$

$\qquad 5y + 7x = 0$

51. $-x + 3y = 5$

$\qquad 3y = x + 5$

$\qquad y = \frac{x}{3} + \frac{5}{3}$

$\qquad m = \frac{1}{3}, \ m' = -3$

$\qquad y - 1 = -3(x+2)$

$\qquad y - 1 = -3x - 6$

$\qquad y + 3x = -5$

53. $Ax + By = C$

$\qquad By = -Ax + C$

$\qquad y = \frac{-A}{B}x + \frac{C}{B}$

$\qquad m = \frac{-A}{B}, \ m' = \frac{B}{A}$

$\qquad y - y_1 = \frac{B}{A}(x - x_1)$

$\qquad Ay - Ay_1 = Bx - Bx_1$

$\qquad Bx_1 - Ay_1 = Bx - Ay$

55. $x+3y=2$

If $x=0$, $y=\frac{2}{3}$

If $y=0$, $x=2$

$\frac{x}{2}+\frac{y}{\frac{2}{3}}=1$

Section 4.2   Mathematical Models

1a. $C=5(300)+400=1900$

1b. $A=\frac{C}{x}=\frac{1900}{300}=6.33$

1c. $A=\frac{5(500)+400}{500}=\frac{2900}{500}=5.80$

1d. $m=5$

1e. $2000=5x+400$

$5x=1600$

$x=\frac{1600}{5}=320$

3a. $m=2$

3b. $v=2(100)=200$

3c. $b=300$, unchanged

3d. $100=2x+300$

$2x=-200$

$x=-100$, No items

$500=2x+300$

$2x=200$

$x=100$

$1000=2x+300$

$2x=700$

$x=350$

5a. $R=ux=30(2810)=84.300$

5b. $P=ux-mx-b=30(2810)-8(2810)-24{,}000=37{,}820$

5c. $P=(30-8)x-24{,}000$

$O=22x-24000$

$X=\frac{24000}{22}=1090.91\simeq1091$

7a. $v=.37x=.37(50{,}000)=18{,}500$

7b. $P=(u-m)x-b$

$=(.62-.37)(50{,}000)-18{,}000=.25(50{,}000)-18{,}000=-5500$ (a loss)

7c. $O=.25x-18{,}000$

$X=\frac{18000}{.25}=72{,}000$

9. $I_1 = \frac{1}{2}$, $v_1 = 6$

$I_2 = \frac{2}{3}$, $v_2 = 8$

$I - \frac{1}{2} = \left(\dfrac{\frac{1}{2} - \frac{2}{3}}{6 - 8} \right)(v - 6)$

$I - \frac{1}{2} = \frac{1}{12}(v - 6)$

$12I - 6 = v - 6$

$I = \frac{v}{12}$

11. $V = C - \dfrac{Cn}{N}$

$V = 300 - \dfrac{300(5)}{20} = 225$

13. $I = Art = 4000(.12)3 = 1440$

$P = \dfrac{4000 + 1440}{35} = 151.11$

Section 4.3   Quadratic Functions: The Parabolic Graph

1. $f(0) = 0 - 4 = -4$

$x^2 - 4 = 0$

$(x - 2)(x + 2) = 0$, $x = \pm 2$

vertex: $(0, -4)$

3. $f(0) = 1$

$x^2 + 1 = 0$, no x-intercepts

vertex: $(0, 1)$

5. $f(0) = 9$

$9 - x^2 = 0$

$(3 - x)(3 + x) = 0$, $x = \pm 3$

vertex: $(0, 9)$

7. $x(0) = 0$

$t^2 - 3t = 0$

$t(t - 3) = 0$, $t = 0, 3$

vertex: $\left(\frac{3}{2}, f(\frac{3}{2})\right) = (\frac{3}{2}, \frac{-9}{4})$

9. $f(0) = 6$

$x^2 - 5x + 6 = 0$

$(x - 3)(x - 2) = 0$, $x = 2, 3$

vertex: $\left(\frac{5}{2}, f(\frac{5}{2})\right) = \left(\frac{5}{2}, \frac{-1}{4}\right)$

11. $f(0) = 1$

$3x^2 + 2x + 1 = 0$

$x = \dfrac{-2 \pm \sqrt{4 - 12}}{6}$

no real roots

vertex: $\left(\frac{-2}{6}, f(\frac{-2}{6})\right) = \left(\frac{-1}{3}, \frac{2}{3}\right)$

13. $f(0) = 2$

$-x^2 + x + 2 = 0$

$x^2 - x - 2 = 0$

$(x - 2)(x + 1) = 0$, $x = -1, 2$

vertex: $\left(\frac{1}{2}, f(\frac{1}{2})\right) = \left(\frac{1}{2}, \frac{9}{4}\right)$

15. $f(0) = -5$

$2x^2 - 3x - 5 = 0$

$(2x - 5)(x + 1) = 0$, $x = -1, \frac{5}{2}$

vertex: $\left(\frac{3}{4}, f(\frac{3}{4})\right) = \left(\frac{3}{4}, \frac{-49}{8}\right)$

17. $f(0)=3$

$3+x-x^2 = 0$

$x^2-x-3 = 0$

$$x = \frac{1\pm\sqrt{1-4(1)(-3)}}{2} = \frac{1\pm\sqrt{13}}{2}$$

vertex: $\left(\frac{1}{2}, f\left(\frac{1}{2}\right)\right) = \left(\frac{1}{2}, \frac{13}{4}\right)$

19. $G(0)=1$

$x^2+x+1 = 0$

$$x = \frac{-1\pm\sqrt{1-4}}{2}, \text{ no real roots}$$

vertex: $\left(\frac{-1}{2}, f\left(\frac{-1}{2}\right)\right) = \left(\frac{-1}{2}, \frac{3}{4}\right)$

21. $y = k(x-0)(x-2)$

$-1 = k(1)(1-2)$

$k = 1$

$y = x(x-2) = x^2-2x$

23. $y = k(x-1)^2$

$1 = k(-1)^2$

$k = 1$

$y = (x-1)^2$

25. $y = k(x+1)(x+3)$

$1 = k(-2+1)(-2+3)$

$k = -1$

$y = -(x+1)(x+3) = -x^2-4x-3$

27. $y = k(x+1)(x-3)$

$-2 = k(1)(-3)$

$k = \frac{2}{3}$

$y = \frac{2}{3}(x+1)(x-3)$

$y = \frac{2x^2}{3} - \frac{4x}{3} - 2$

29. $y = k(x-2)(x+5)$

$1 = k(-2)(5)$

$k = -\frac{1}{10}$

$y = -\frac{1}{10}(x-2)(x+5)$

$= -\frac{1}{10}x^2 - \frac{3x}{10} + 1$

35 b. $P(t) = 45t-3t^2$

$45t-3t^2 > 0$

$3t(15-t) > 0$

c. $P(t)<0$ for $15<t<24$

37. $v = 100+200t-25t^2$

vertex: $\left(\frac{-200}{-50}, v(4)\right) = (4{,}500)$

$v_{max} = 500$

$100+200t-25t^2 = 0$

$$t = \frac{-200\pm\sqrt{200^2-4(-25)(100)}}{2(-25)} = 4\pm4.47$$

$t = 8.47$

39. $f \circ g = f(ax^2 + bx + c) = A(ax^2 + bx + c) + B(ax^2 + bx + c) + C$

$\qquad = A\left(a^2x^4 + 2abx^3 + (2ac + b)x^2 + 2bcx + c^2\right) + B(ax^2 + bx + c) + C \quad$ is not a quadratic

Section 4.4   Polynomial Functions

1. $y = x^3 + 1 = (x+1)(x^2 - x + 1) = 0,\ x = -1$

$\quad f(0) = 1$

$$\begin{array}{c} -\ +\quad +\ + \\ \underline{\qquad\quad+\qquad\quad} \\ -1 \end{array}$$

3. $y = 1 - x^4 = (1 + x^2)(1 - x)(1 + x) = 0,\ x = \pm 1$

$\quad f(0) = 1$

$$\begin{array}{c} +\ +\ -\quad +\ +\ +\quad +\ -\ + \\ \underline{\qquad+\qquad\qquad+\qquad} \\ -1 \qquad\quad 1 \end{array}$$

5. $y - 3 = (x-1)^3$

$\quad$ Translate $(0,\,0)$ of $y = x^3$ to $(1,\,3)$

7. $y + 2 = (x+1)^5$

$\quad$ Translate $(0,\,0)$ of $y = x^5$ to $(-1,\,-2)$

9. $y - 1 = -(x-2)^4$

$\quad$ Translate $(0,\,0)$ to $y - -x^4$ to $(2,\,1)$

11. $y = x(x-1)(x+1) = 0,\ x = 0,\ \pm 1$

$\quad f(0) = 0$

$$\begin{array}{c} -\ -\ -\quad -\ -\ +\quad +\ -\ +\quad +\ +\ + \\ \underline{\qquad+\qquad\quad+\qquad\quad+\qquad} \\ -1 \qquad 0 \qquad 1 \end{array}$$

13. $y = x^2(x-1)(x+1) = 0,\ x = 0,\ \pm 1$

$\quad f(0) = 0$

$$\begin{array}{c} +\ -\ -\quad +\ -\ +\quad +\ -\ +\quad +\ +\ + \\ \underline{\qquad+\qquad\quad+\qquad\quad+\qquad} \\ -1 \qquad 0 \qquad 1 \end{array}$$

15. $y = x^3(x-2)(x+2) = 0,\ x = 0,\ \pm 2$

$\quad f(0) = 0$

$$\begin{array}{c} -\ -\ -\quad -\ -\ +\quad +\ -\ +\quad +\ +\ + \\ \underline{\qquad+\qquad\quad+\qquad\quad+\qquad} \\ -2 \qquad 0 \qquad 2 \end{array}$$

17. $y = 3(x^3 - 27) = 3(x-3)(x^2 + 3x + 9) = 0$, $x = 3$

$f(0) = -81$

```
        + - +   + + +
    ————————————|——————————
               3
```

19. $y = (x^2 - 1)^2 = (x-1)^2(x+1)^2 = 0$, $x = \pm 1$

$f(0) = 1$

```
     + +    + +    + +
    ——————|————————|——————
         -1        1
```

21. $y = x(x-1)(x+3) = 0$, $x = 0, 1, -3$

$f(0) = 0$

```
    — — —    — — +    + — +    + + +
    ——————|————————|————————|——————
         -3        0        1
```

23. $y = (x+1)^2 x(x-1) = 0$, $x = -1, 0, 1$

$f(0) = 0$

```
    + — —    + — —    + + —    + + +
    ——————|————————|————————|——————
         -1        0        1
```

25. $y = (x^2 - 9)(x^2 + 4) = (x-3)(x+3)(x^2 + 4) = 0$, $x = \pm 3$

$f(0) = (-9)(4) = -36$

```
      — — +   — + +   + + +
    ————————|————————|——————
           -3        3
```

27. $y = (x+2)(x-3)(x+5) = 0$, $x = -2, 3, -5$

$f(0) = -30$

```
      — — —   — — +   + — +   + + +
    ————————|————————|————————|——————
           -5       -2        3
```

29. $y = (x-5)(x-1)(x+2)^2 = 0$, $x = 5, 1, -2$

$f(0) = (-5)(-1)(4) = 20$

```
      — — +   — — +   — + +   + + +
    ————————|————————|————————|——————
           -2        1        5
```

3. $p = \dfrac{3x - 2}{x - 4}$

Vert. A.: $x - 4 = 0$, $x = 4$

Horiz. A.: $y = \dfrac{3 - \frac{2}{x}}{1 - \frac{4}{x}}$, as $x \to \infty$, $y \to \dfrac{3}{1}$, $y = 3$

Intercepts: $f(0) = \dfrac{-2}{-4} = \dfrac{1}{2}$, $\left(0, \dfrac{1}{2}\right)$

$3x - 2 = 0$, $x = \dfrac{2}{3}$, $\left(\dfrac{2}{3}, 0\right)$

5. $p = \dfrac{1}{w + 1}$

Vert. A.: $w + 1 = 0$, $w = -1$

Horiz. A.: $p = \dfrac{\frac{1}{w}}{1 + \frac{1}{w}}$, as $w \to \infty$, $p = 0$

Intercepts: $p(0) = 1$, $(0, 1)$

$p = 0 = \dfrac{1}{w + 1}$, no w-intercept

7. $s = \dfrac{5w^2}{4 - w^2}$

Vert. A.: $4 - w^2 = (2 - w)(2 + w) = 0$, $w = \pm 2$

Horiz. A.: $s = \dfrac{5}{\frac{4}{w^2} - 1}$, as $w \to \infty$, $s = -5$

Intercepts: $s(0) = 0$, $(0, 0)$

$0 = \dfrac{5w^2}{4 - w^2}$, $(0, 0)$

9. $y = \dfrac{x^2}{x^2 - 9}$

Vert. A.: $x^2 - 9 = (x-3)(x+3)=0$, $x = \pm 3$

Horiz. A.: $y = \dfrac{1}{1 - \dfrac{9}{x^2}}$, as $x \to \infty$, $y = 1$

Intercepts: $y(0) = 0$, $(0, 0)$

$$x^2 = 0,\ x = 0,\ (0, 0)$$

11. $s = \dfrac{(t+2)^2}{t(t+2)} = \dfrac{t^2 + 4t + 4}{t^2 + 2t} = \dfrac{t + 2}{t}$

Vert. A.: $t = 0$

Horiz. A.: $s = \dfrac{1 + \frac{4}{t} + \frac{4}{t^2}}{1 + \frac{2}{t}}$, as $t \to \infty$, $s = 1$

Intercepts: $s(0)$ does not exist

$$s = \dfrac{(t+2)^2}{t(t+2)} = 0,\ t = -2,\ s(-2) \text{ does not exist}$$

13. $y = \dfrac{3}{w^2 - 9w} = \dfrac{3}{w(w-9)}$

Vert. A.: $w(w-9) = 0$, $w = 0, 9$

Horiz. A.: $y = \dfrac{\frac{3}{w^2}}{1 - \frac{9}{w}}$, as $w \to \infty$, $y = 0$

Intercepts: $y(0)$ does not exist

$$y = 0 = \dfrac{3}{w(w-9)}, \text{ no solution}$$

15. $y = \dfrac{1}{(p-4)(p+1)}$

Vert. A.: $(p-4)(p+1) = 0$, $p = -1, 4$

Horiz. A.: $y = \dfrac{\frac{1}{p^2}}{1 - \frac{3}{p} - \frac{4}{p^2}}$, as $p \to \infty$, $y = 0$

Intercepts: $y(0) = \dfrac{1}{(-4)(1)} = -\dfrac{1}{4}$

$$y = 0 = \dfrac{1}{(p-4)(p+1)}, \text{ no solution}$$

17. $y = \dfrac{5x + 3}{x(x+1)}$

Vert. A.: $x(x+1) = 0$, $x = 0, -1$

Horiz. A.: $y = \dfrac{\dfrac{5}{x} + \dfrac{3}{x^2}}{1 + \dfrac{1}{x}}$, as $x \to \infty$, $y = 0$

Intercepts: $y(0) = $ does not exist

$$y = 0 = \frac{5x + 3}{x(x+1)}, \quad 0 = 5x + 3, \quad x = -\frac{3}{5}$$

21. $u = \dfrac{v^2 + 3}{4v^2 + 5}$

Vert. A.: $4v^2 + 5 = 0$, no solution

Horiz. A.: $u = \dfrac{1 + \dfrac{3}{v^2}}{4 + \dfrac{5}{v^2}} = \dfrac{1}{4}$

Intercepts: $u(0) = \dfrac{3}{5}$

$$u = 0 = \frac{v^2 + 3}{4v^2 + 5}, \text{ no solution}$$

Section 4.6   Multipart Functions

17. $f(x) = |x+1| = \begin{cases} x+1, & x \geq -1 \\ -(x+1), & x < -1 \end{cases}$

19. $H(x) = |2x| = \begin{cases} 2x, & x \geq 0 \\ -2x, & x < 0 \end{cases}$

21. $f(x) = |x| - 3 = \begin{cases} x - 3, & x \geq 0 \\ -x - 3, & x < 0 \end{cases}$

23. $y = |x| + 3 = \begin{cases} x + 3, & x \geq 0 \\ -x + 3, & x < 0 \end{cases}$

25. $f(t) = |t^2| = t^2$

27. $f(x) = x^2 - |x| + 1 = \begin{cases} x^2 - x + 1, & x \geq 0 \\ x^2 + x + 1, & x < 0 \end{cases}$

29. $g(x) = |2x^2 + x - 1| = |(2x-1)(x+1)|$

$(2x-1)(x+1) = 0$, $x = -1, \frac{1}{2}$

```
2x-1       -     -     +
x+1        -     +     +
       ----+-----+----
          -1     1/2
```

$$g(x) = \begin{cases} 2x^2 + x - 1 & x \leq -1 \\ -(2x^2 + x - 1) & -1 < x < \frac{1}{2} \\ 2x^2 + x - 1 & x \geq \frac{1}{2} \end{cases}$$

31. $y = x^2 + 3|x| + 2 = \begin{cases} x^2 + 3x + 2 & x \geq 0 \\ x^2 - 3x + 2 & x < 0 \end{cases}$

33. $f(x) = (x-1)|x+1| = \begin{cases} (x-1)(x+1), & x \geq 1 \\ -(x-1)(x+1), & x < -1 \end{cases}$

35. $f(x) = x|x-4| = \begin{cases} x(x-4), & x \geq 4 \\ -x(x-4), & x < 4 \end{cases}$

Chapter 4   Review

1. $m = \dfrac{3-(-2)}{-1-0} = -5$

3. $m = \dfrac{2-(-5)}{3-1} = \dfrac{7}{2}$

7. $m = \dfrac{4-(-1)}{5-(-1)} = \dfrac{5}{6}, \; m' = -\dfrac{6}{5}$

9. $y - (-1) = -2(x-3)$
$$y + 1 = -2x + 6$$
$$y + 2x = 5$$

11. $y - 2 = \left(\dfrac{2-4}{7-(-1)}\right)(x-7)$
$$y - 2 = -\dfrac{1}{4}(x-7)$$
$$4y - 8 = -x + 7$$
$$4y + x = 15$$

13. $2x + 7y = -3$
$$7y = -2x - 3$$
$$y = -\dfrac{2}{7}x - \dfrac{3}{7}$$
$$m = -\dfrac{2}{7}x, \; b = -\dfrac{3}{7}$$

15. $3x - 6y = 1$
$$6y = 3x - 1$$
$$y = \dfrac{1}{2}x - \dfrac{1}{6}$$
$$m = \dfrac{1}{2}, \; b = -\dfrac{1}{6}$$

17. $x - 2y = -5$
$$2y = x + 5$$
$$y = \dfrac{1}{2}x + \dfrac{5}{2}$$
$$m = \dfrac{1}{2}, \; b = \dfrac{5}{2}, \; m' = -2$$
$$y - \dfrac{5}{2} = -2(x-0)$$
$$2y - 5 = -4x$$
$$2y + 4x = 5$$

19. $x - 3y = 7$
$$3y = x - 7$$
$$y = \dfrac{1}{3}x - \dfrac{7}{3}$$
$$m = \dfrac{1}{3}, \; b = -\dfrac{7}{3}, \; m' = -3$$
$$y - (-2) = -3(x-1)$$
$$y + 2 = -3x + 3$$
$$y + 3x = 1$$

21. $y = x^2 + 7$
$y(0) = 7$, vertex: $(0, 7)$

23. $f(x)=2x^2-x-3$

$f(0)=-3$

$(2x-3)(x+1)=0,\ x=\frac{3}{2},\ -1$

vertex: $\left(\frac{1}{4},\ f\left(\frac{1}{4}\right)\right)=\left(\frac{1}{4},\ \frac{-25}{8}\right)$

25. $s=2+t-t^2$

$s(0)=2$

$2+t-t^2=(2-t)(1+t)=0,\ t=-1,\ 2$

vertex: $\left(\frac{1}{2},\ f\left(\frac{1}{2}\right)\right)=\left(\frac{1}{2},\ \frac{9}{4}\right)$

27. x-intercepts: $(-2,\ 0),\ (6,\ 0)$

$y=k(x+2)(x-6)$

y-intercepts: $(0,\ 1)$

$1=k(2)(-6)$

$k=-\frac{1}{12}$

$y=-\frac{1}{12}(x+2)(x-6)$

29. x-intercepts: $(-1,\ 0)(-3,\ 0)$

$y=k(x+1)(x+3)$

y-intercept: $(0,\ 2)$

$2=k(1)(+3)$

$k=\frac{2}{3}$

$y=\frac{2}{3}(x+1)(x+3)$

31. $y=2-x^3=0,\ x^3=2,\ x=\sqrt[3]{2}$

$f(0)=2$

```
    +        −
  ──────┼──────
       ³√2
```

33. $y=x(x+1)(x+5)=0,\ x=0,\ -1,\ 5$

$f(0)=0$

```
 − − −    − − +    − + +    + + +
 ───────┼────────┼────────┼───────
       −5       −1        0
```

35. $y=(x-2)(x+1)^2=0,\ x=2,\ -1$

$f(0)=(-2)(1)=-2$

```
      − +    − +    + +
    ──────┼──────┼──────
         −1      2
```

37. $y=\frac{3}{x+2}$

Vert. A: $x+2=0,\ x=-2$

Horiz. A: $y=\frac{3}{x+2}=0$, no solution

Intercepts: $y(0)=\frac{3}{2},\ \left(0,\ \frac{3}{2}\right)$

$0=\frac{3}{x+2}$, no solution

39. $p = \dfrac{2x}{(x-2)(x+2)}$

Vert. A: $(x-2)(x+2) = 0$, $x = \pm 2$

Horiz. A: $p = \dfrac{\frac{2}{x}}{1 - \frac{4}{x^2}}$, as $x \to \infty$, $p = 0$

Intercepts: $p(0) = 0$, $(0, 0)$

$$0 = \dfrac{2x}{(x-2)(x+2)}, \ x = 0, \ (0, 0)$$

45. $F(x) = \left| x - \dfrac{1}{2} \right| = \begin{cases} x - \dfrac{1}{2}, & x \geq \dfrac{1}{2} \\ -x + \dfrac{1}{2}, & x < \dfrac{1}{2} \end{cases}$

47. $y = |t| + 2 = \begin{cases} t + 2, & t \geq 0 \\ -t + 2, & t < 0 \end{cases}$

49. $s = t^2 + 2t = t(t+2) = 0$, $t = 0, -2$
$s(0) = 0$
vertex: $\left(-\dfrac{2}{2}, f(-1) \right) = (-1, -1)$

51. $y = x^2 - x = x(x-1) = 0$, $x = 0, 1$
$y(0) = 0$
vertex: $\left(\dfrac{1}{2}, f(\dfrac{1}{2}) \right) = \left(\dfrac{1}{2}, -\dfrac{1}{4} \right)$

57. $v - v_1 = \left(\dfrac{v_1 - v_2}{t_1 - t_2} \right)(t - t_1)$

$v - 25 = \left(\dfrac{25 - 38}{4 - 10} \right)(t - 4)$

$v - 25 = \dfrac{13}{6}(t - 4)$

$v = \dfrac{13t}{6} + \dfrac{98}{6}$

59. $m = \dfrac{-315 - (-225)}{0 - 500} = \dfrac{9}{50}$

61. $P(t) = 10(2t - 3) = 20t - 30$

$P(t) = 20t - 30 > 0$

$20t > 30$, $t > \dfrac{3}{2}$

63a. $P(n) = 12n - 5500 > 0$

63c. $P(n) = 12n - 5500 > 0$

$12n > 5500$, $n > 459$

Chapter 5 Exponential and Logarithmic Functions

Section 5.1   Exponential Functions

1. $x^{-3/2} = \dfrac{27}{8}$

$\left(x^{-3/2}\right)^{-2/3} = \left(\dfrac{27}{8}\right)^{-2/3}$

$x = \left(\dfrac{3}{2}\right)^{-2} = \dfrac{4}{9}$

3. $27^{x+2} = 9^{2-x}$

$(3^3)^{x+2} = (3^2)^{2-x}$

$3(x+2) = 2(2-x)$

$3x+6 = 4-2x$

$5x = -2$

$x = -\dfrac{2}{5}$

5. $(-2)^{x+2} = -8$

$(-2)^{x+2} = (-2)^3$

$x+2 = 3$

$x = 1$

7. $\left((3)^x\right)^x = 1$

$3^{x^2} = 1$

$x^2 = 0$

$x = 0$

9. $3^{x^2+4x} = \dfrac{1}{81}$

$3^{x^2+4x} = (3)^{-4}$

$x^2 + 4x = -4$

$x^2 + 4x + 4 = 0,\ (x+2)^2 = 0,\ x = -2$

41. $S = 500{,}000(.98)^t$

$S(4) = 500{,}000(.98)^4 = 461{,}184$

43. $I = P(1+i)^t - P$

$= 500(1.045)^5 - 500 = 123.09$

45. $P = 1000(1+.035)^{10} = 1410.60$

47. $I = 95{,}000(1+.015)^8 - 5000 = 632.46$

Section 5.2   Exponential Functions with Base e

11. $v = 10.772e^{(2.032)(2.75)} = 10.772e^{5.588} = 10.772(267.2) = 2878.29$

13. $T = 198.6 + 95.3e^{(-.147)(7.5)} = 198.6 + 95.3e^{-1.1025} = 198.6 + 95.3(.3320) = 230.24$

17. $f(t) = \sqrt{t+3.346}\,e^{.0178t}$

$f(30.542) = \sqrt{30.542+3.346}\,e^{(0.178)(30.542)}$

$= \sqrt{33.888}\,e^{.54365} = (5.8213)(1.7223) = 10.026$

19. $H(.7683) = \left(e^{(.15)(.7683)} + e^{(-.15)(.7683)}\right)e^{3.19-.7683}$

$= \left(e^{.11525} + e^{-.11525}\right)e^{2.4217}$

$= (1.122+.8911)(11.265) = 22.68$

Section 5.3   The Logarithm Function

7. $\log_b 4 = 2$
   $b^2 = 4,\ b = 2$

9. $\log_b 100 = 2$
   $b^2 = 100,\ b = 10$

11. $\log_{10} x = 4$
    $10^4 = x$

13. $\log_x 10 = 1$
    $x^1 = 10$

15. $\log_x 64 = 3$
    $x^3 = 64$
    $x = 4$

17. $\log_{27} x = \frac{2}{3}$
    $27^{2/3} = x$
    $x = 3^2 = 9$

19. $\log_3 9 = x$
    $3^x = 9 = 3^2$
    $x = 2$

21. $\log_b b^a = x$
    $b^x = b^a$
    $x = a$

23. $\log_x 2 = \frac{1}{3}$
    $x^{1/3} = 2$
    $x = 2^3 = 8$

25. $\log_x 6 = \frac{1}{2}$
    $x^{1/2} = 6$
    $x = 6^2 = 36$

29. $\log_2 (x+3) = -1$
    $2^{-1} = x+3$
    $x = \frac{1}{2} - 3 = -\frac{5}{2}$

31. $\log_3 (x+1) < 2$
    $3^2 > x+1$
    $8 > x$
    $x+1 > 0,\ x > 1$
    $1 < x < 8$

33. $2 \leq \log_2 x \leq 3$

    $2 \leq \log_2 x \quad \log_2 x \leq 3$

    $2^2 \leq x \quad\quad\quad x \leq 2^3$

                  $4 \leq x \leq 8$

51. $2 = \log_a 4$

    $a^2 = 4, \ a = 2$

53. $-1 = \log_a {}^{(.1)}$

    $a^{-1} = \frac{1}{10}$

       $a = 10$

57a. $f(x) = \log_2 x$

    $f(x+y) = \log_2 (8+8) = \log_2 16 = 4$

    $f(x) + f(y) = \log_2 8 + \log_2 8 = 3+3$

    $f(x) + f(y) \neq f(x+y)$

57b. $f(ax) = f(4 \cdot 2) = \log_2 (4 \cdot 2) = 3$

    $af(x) = 4f(2) = 4 \log_2 2 = 4$

    $f(ax) \neq af(x)$

69. $A(t) = A_0 e^{-t/5}$

    $\frac{A_0}{2} = A_0 e^{-t/5}$

    $\frac{1}{2} = e^{-t/5}$

    $\ln \frac{1}{2} = \ln e^{-t/5}$

    $\ln \frac{1}{2} = \frac{-t}{5}$

    $t = -5 \ln \frac{1}{2} = 3.4657$

Section 5.4   Basic Properties of Logarithms

1. $\log_2 32 \cdot 16 = \log_2 32 + \log_2 16 = \log_2 2^5 + \log_2 2^4 = 5 + 4 = 9$

3. $\log_5 25^{1/4} = \frac{1}{4} \log 5^2 = \frac{1}{4} \cdot 2 = \frac{1}{2}$

5. $\log_3 27 \cdot 9 \cdot 3 = \log_3 3^3 + \log_3 3^2 + \log_3 3 = 3 + 2 + 1 = 6$

7. $\log_2 (8 \cdot 32)^3 = 3\left[\log_2 8 + \log_2 32\right] = 3 \log_2 2^3 + 3 \log_2 2^5 = 9 + 15 = 24$

9. $\log \frac{3}{2} = \log 3 - \log 2 = .1761$

11. $\log 12 = \log 3 \cdot 4 = \log 3 + \log 4 = \log 3 + \log 2^2 = 1.0791$

13. $\log 90 = \log 9 \cdot 10 = \log 9 + \log 10 = \log 3^2 + \log 10 = 1.9542$

15. $\log \sqrt{5} = \log \left(\frac{10}{2}\right)^{1/2} = \frac{1}{2}\left[\log 10 - \log 2\right] = .3495$

17. $\log 2400 = \log 8 \cdot 3 \cdot 10^2 = \log 8 + \log 3 + \log 10^2 = 3.3801$

19. $\log .0014 = \log 2 \cdot 7 \cdot 10^{-4} = \log 2 + \log 7 + \log 10^{-4} = -2.8539$

21. $\log_2 x^2 - \log_2 x = \log_2 \frac{x^2}{x} = \log_2 x$

   or

   $\log_2 x^2 - \log_2 x = 2 \log_2 x - \log_2 x = \log_2 x$

23. $\log x + \log \frac{1}{x} = \log x \cdot \frac{1}{x} = \log 1 = 0$

25. $\log 5t + 2 \log (t^2 - 4) - \frac{1}{2} \log (t+3)$

   $= \log \dfrac{5t(t^2-4)^2}{(t+3)^{1/2}}$

27. $3 \log u - 2 \log (u+1) - 5 \log (u-1)$

   $= \log \dfrac{u^3}{(u+1)^2(u-1)^5}$

29. $\qquad \ln I = \left(\frac{-R}{L}\right)t + \ln I_0$

   $\ln I - \ln I_0 = \dfrac{-Rt}{L}$

   $\ln \dfrac{I}{I_0} = \dfrac{-Rt}{L}, \; \dfrac{I}{I_0} = e^{-\frac{R}{L}t}$

Section 5.5   Exponential and Logarithmic Equations

1.
$$7^{x+1} = 2^x$$
$$\log 7^{x+1} = \log 2^x$$
$$(x+1) \log 7 = x \log 2$$
$$x \log 7 + \log 7 = x \log 2$$
$$x(\log 7 - \log 2) = -\log 7$$
$$x = \frac{-\log 7}{\log 7 - \log 2} = -1.553$$

3.
$$10^{x^2} = 2^x$$
$$\log 10^{x^2} = \log 2^x$$
$$x^2 \log 10 = x \log 2$$
$$x^2 - x \log 2 = 0$$
$$x(x - \log 2) = 0$$
$$x = 0, \; \log 2 = .301$$

– 64 –

5.
$$2^{x+1} = 3$$
$$\log 2^{x+1} = \log 3$$
$$(x+1) \log 2 = \log 3$$
$$x \log 2 + \log 2 = \log 3$$
$$x \log 2 = \log 3 - \log 2$$
$$x = \frac{\log 3 - \log 2}{\log 2} = .5850$$

7.
$$2^{x+5} = 3^{x-2}$$
$$\log 2^{x+5} = \log 3^{x-2}$$
$$(x+5) \log 2 = (x-2) \log 3$$
$$x(\log 2 - \log 3) = -2 \log 3 - 5 \log 2$$
$$x = \frac{-2 \log 3 - 5 \log 2}{\log 2 - \log 3}$$
$$= 13.97$$

9.
$$2^{x+1} > 5$$
$$\ln 2^{x+1} > \ln 5$$
$$(x+1) \ln 2 > \ln 5$$
$$x \ln 2 > \ln 5 - \ln 2$$
$$x > \frac{\ln 5 - \ln 2}{\ln 2} = 1.32$$

11. $2^{\log x} = 2$
$$\log x = 1$$
$$x = 10$$

13.
$$(\log x)^{1/2} = \log \sqrt{x} = \log x^{1/2}$$
$$(\log x)^{1/2} = \frac{1}{2} \log x$$
$$\log x = \frac{1}{4} \log^2 x$$
$$\frac{1}{4} \log^2 x - \log x = 0$$
$$\log x \left(\frac{1}{4} \log x - 1\right) = 0$$
$$\log x = 0 \qquad \frac{1}{4} \log x = 1$$
$$x = 1 \qquad \log x = 4$$
$$x = 10^4$$

15.
$$(\log x)^3 = \log x^3 = 3 \log x$$
$$(\log x)^3 - 3 \log x = 0$$
$$\log x (\log^2 x - 3) = 0$$
$$\log x = 0, \qquad \log x = \sqrt{3}, \qquad \log x = -\sqrt{3}$$
$$x = 1 \qquad x = 10^{\sqrt{3}} \qquad x = 10^{-\sqrt{3}}$$

17. $\log (x+15) + \log x = 2$

$\log x (x+15) = 2$

$x (x+15) = 10^2$

$x^2+15x-100 = 0$

$(x+20)(x-5) = 0$

$x= 5$, $x= -20$, impossible, $\log (-20)$ is not defined

19. $\log (x-2) - \log (2x+1) = \log \frac{1}{x}$

$\log \frac{(x-2)}{(2x+1)} = \log \frac{1}{x}$

$\frac{x-2}{2x+1} = \frac{1}{x}$

$x(x-2) = 2x+1$

$x^2-4x-1 = 0$

$x = 2+\sqrt{5}$, $x= 2-\sqrt{5}$, impossible, $\log (2-\sqrt{5})$ is not defined

21. $\log (x+2) - \log (x-2) - \log x + \log (x-3) = 0$

$\log \frac{(x+2)(x-3)}{x(x-2)} = 0$

$\frac{(x+2)(x-3)}{x(x-2)} = 10^0 = 1$

$x^2 - 2x = x^2 - x - 6$

$x = 6$

23. $\log x + \log (x^2-4) - \log 2x = 0$

$\log \frac{x(x^2-4)}{2x} = 0$

$\frac{x^2-4}{2} = 10$

$x^2-6 = 0$

$x = \sqrt{6}$, $x= -\sqrt{6}$, impossible, $\log (-\sqrt{6})$ is not defined

25. $\log\ (x+1) - \log\ x < 1$

$$\log\left(\tfrac{x+1}{x}\right) < 1$$

$$\tfrac{x+1}{x} < 10^1$$

$$x+1 < 10x$$

$$\tfrac{1}{9} < x$$

27. $\cosh\ x = 2$

$$\frac{e^x + e^{-x}}{2} = 2$$

$$e^x + e^{-x} = 4$$

$$e^x + \tfrac{1}{e^x} = 4$$

$$e^{2x} + 1 = 4e^x$$

$$e^{2x} - 4e^x + 1 = 0$$

$$e^x = \frac{4 \pm \sqrt{16-4}}{2} = 2 \pm \sqrt{3}$$

$$x = \ln\ (2+\sqrt{3})$$

$$x = \ln\ (2-\sqrt{3})$$

31. $A(t) = A_0 e^{-kt}$

$$\frac{A_0}{2} = A_0 e^{-k(28)}$$

$$\tfrac{1}{2} = e^{-28k}$$

$$\ln\ \tfrac{1}{2} = \ln\ e^{-28k} = -28k\ (\ln\ e)$$

$$\ln\ \tfrac{1}{2} = -28k$$

$$k = -\tfrac{1}{28}\ \ln\ \tfrac{1}{2} = .02476$$

33. $P = 50e^{-t/250}$

$$25 = 50e^{-t/250}$$

$$.5 = e^{-t/250}$$

$$\ln\ .5 = \ln\ e^{-t/250}$$

$$\ln\ .5 = \frac{-t}{250}$$

$$t = -250\ \ln\ .5 = 173.3$$

35. $P = P_0 e^{kt}$

$$2P_0 = P_0 e^{.0132t}$$

$$2 = e^{.0132t}$$

$$\ln\ 2 = \ln\ e^{.0132t}$$

$$\ln\ 2 = .0132t$$

$$t = \frac{1}{.0132}\ \ln\ 2 = 52.5$$

Chapter 5   Review

1.

$$2^{x-5} = 3^x$$

$$\ln\ 2^{x-5} = \ln\ 3^x$$

$$(x-5)\ \ln\ 2 = x\ \ln\ 3$$

$$x(\ln\ 2 - \ln\ 3) = 5\ \ln\ 2$$

$$x = \frac{5\ \ln\ 2}{\ln\ 2 - \ln\ 3} = -8.55$$

3.

$$2^{x^2-3x} = 16$$

$$2^{x^2-3x} = 2^4$$

$$x^2 - 3x = 4$$

$$x^2 - 3x - 4 = 0$$

$$(x-4)(x+1) = 0$$

$$x = -1, 4$$

5. $\quad 3^{1-x} = 5$

$\quad\quad \ln 3^{1-x} = \ln 5$

$\quad\quad (1-x) \ln 3 = \ln 5$

$\quad\quad x \ln 3 = \ln 3 - \ln 5$

$\quad\quad x = \dfrac{\ln 3 - \ln 5}{\ln 3} = -.465$

7. $\quad \log x = 3,$

$\quad\quad x = 10^3$

9. $\quad \log_3 81 = x$

$\quad\quad 3^x = 81 = 3^4$

$\quad\quad x = 4$

11. $\quad \log_x 64 = 6$

$\quad\quad x^6 = 64$

$\quad\quad x^6 = 2^6$

$\quad\quad x = 2$

13. $\quad x^{4.3} = 2.1$

$\quad\quad \ln x^{4.3} = \ln 2.1$

$\quad\quad 4.3 \ln x = \ln 2.1$

$\quad\quad \ln x = \dfrac{1}{4.3} \ln 2.1$

$\quad\quad x = 1.188$

25. $\quad \log x + \log x^2 = \log x^3 = 3 \log x$

27. $\quad \log_2 2x + 3 \log_2 x = \log_2 2x + \log_2 x^3 = \log_2 2x^4$

29. $\quad 3 \log_5 x - \log_5 (2x-3) = \log_5 x^3 - \log_5 (2x-3) = \log_5 \dfrac{x^3}{2x-3}$

31. $\quad 2^{x+2} = 10$

$\quad\quad \log 2^{x+2} = \log 10$

$\quad\quad (x+2) \log 2 = 1$

$\quad\quad x \log 2 = 1 - 2 \log 2$

$\quad\quad x = \dfrac{1 - 2 \log 2}{\log 2} = 1.322$

33. $\quad 3^{x+1} = 4^x$

$\quad\quad \ln 3^{x+1} = \ln 4^x$

$\quad\quad (x+1) \ln 3 = x \ln 4$

$\quad\quad x (\ln 3 - \ln 4) = -\ln 3$

$\quad\quad x = \dfrac{-\ln 3}{\ln 3 - \ln 4} = 3.819$

35. $\quad \log (\log x) = 1$

$\quad\quad 10^1 = \log x$

$\quad\quad 10^{10} = \log x$

37. $\quad \log x + \log (x+1) = \log 6$

$\quad\quad \log x(x+1) = \log 6$

$\quad\quad x^2 + x = 6$

$\quad\quad x^2 + x - 6 = 0$

$\quad\quad (x+3)(x-2) = 0$

$\quad\quad x = 2, \ x = -3, \log (-3)$ is not defined

39. $\log x + \log (x-3) = 1$

$$\log x(x-3) = 1$$

$$x^2 - 3x = 10$$

$$x^2 - 3x - 10 = 0$$

$$(x-5)(x+2) = 0$$

$x = 5, \cancel{-2}, \log (-2)$ is not defined

41. $A(t) = A_0 e^{-t/4}$

$$\frac{A_0}{2} = A_0 e^{-t/4}$$

$$\ln \frac{1}{2} = \ln e^{-t/4}$$

$$\ln \frac{1}{2} = \frac{-t}{4}$$

$$t = -4 \ln \frac{1}{2} = 2.77$$

45. $D(t) = D(0) e^{-rt/V}$

$r = \frac{1.4L}{\text{min}}$, $V = 450\text{mL} = .45\text{L}$

$t = 5$ sec. $= .0833$ min, $D(0) = 2.3\text{mg}$

$$D(.0833) = 2.3 e^{\frac{-1.4(.0833)}{.45}}$$

$$= 1.77$$

– 69 –

Chapter 6 Trigonometry

<u>Section 6.1 Angles and Their Measurement</u>

1.

3.

5.

7.

9.

11.

13.

$1485°$

15.

$(69°55')$

$-290°5$

17.

$135°$

$(-1,1)$

19.

$-90°$

$(0,-3)$

21.

$180°$

$(-4,0)$

23. $75° \cdot \frac{\pi}{180°} = \frac{5\pi}{12}$

25. $480° \cdot \frac{\pi}{180°} = \frac{8\pi}{3}$

27. $-240° \cdot \frac{\pi}{180°} = \frac{-4\pi}{3}$

29. $95° \cdot \frac{\pi}{180°} = \frac{19\pi}{36}$

31. $750° \cdot \frac{\pi}{180°} = \frac{25\pi}{6}$

41.

$4 = 229.18°$

$(4 - 2\pi = -2.28)$

43.

$\pi = 180°$

$(+\pi)$

45.

$-3\pi = -540°$

47.

$10 = 572.96°$

$(10 - 4\pi = -2.566)$

49.

$100\pi = 18,000°$

51.

$s = r\theta$

$s = 5 \cdot 30° \cdot \dfrac{\pi}{180°}$

$= 5\dfrac{\pi}{6} = 2.62$

53.

$s = r\theta$

$s = 8(81.5°)\dfrac{\pi}{180} = 11.38$

55. $s = r\theta$

$5.2 = r(70.7)\dfrac{\pi}{180}$

$= \dfrac{(5.2)(180)}{(70.7)\pi} = 4.2$

1. $r = \sqrt{2^2 + 4} = 2\sqrt{5}$

    $\sin\theta = \dfrac{2}{\sqrt{5}}$, $\cos\theta = \dfrac{1}{\sqrt{5}}$, $\tan\theta = 2$

    $\csc\theta = \dfrac{\sqrt{5}}{2}$, $\sec\theta = \sqrt{5}$, $\cot\theta = \dfrac{1}{2}$

3. $r = \sqrt{(-9)^2 + 16^2} = \sqrt{337}$

    $\sin\theta = \dfrac{16}{\sqrt{337}}$, $\cos\theta = \dfrac{-9}{\sqrt{337}}$, $\tan\theta = \dfrac{-16}{9}$

    $\csc\theta = \dfrac{\sqrt{337}}{16}$, $\sec\theta = \dfrac{-\sqrt{337}}{9}$, $\cot\theta = \dfrac{-9}{16}$

5. $r = \sqrt{2^2 + (-7)^2} = \sqrt{53}$

    $\sin\theta = \dfrac{-7}{\sqrt{53}}$, $\cos\theta = \dfrac{2}{\sqrt{53}}$, $\tan\theta = \dfrac{-7}{2}$

    $\csc\theta = \dfrac{-\sqrt{53}}{7}$, $\sec\theta = \dfrac{\sqrt{53}}{2}$, $\cot\theta = \dfrac{-2}{7}$

7. $r = \sqrt{3^2 + (-1)^2} = \sqrt{10}$

    $\sin\theta = \dfrac{-1}{\sqrt{10}}$, $\cos\theta = \dfrac{3}{\sqrt{10}}$, $\tan\theta = \dfrac{-1}{3}$

    $\csc\theta = -\sqrt{10}$, $\sec\theta = \dfrac{\sqrt{10}}{3}$, $\cot\theta = -3$

9. $r = \sqrt{(-\sqrt{3})^2 + (-1)^2} = \sqrt{4} = 2$

    $\sin\theta = \dfrac{-1}{2}$, $\cos\theta = \dfrac{-\sqrt{3}}{2}$, $\tan\theta = \dfrac{-1}{\sqrt{3}}$

    $\csc\theta = -2$ $\sec\theta = \dfrac{-2}{\sqrt{3}}$, $\cot\theta = \sqrt{3}$

19. $\tan\theta = \dfrac{3}{4}$, Q I

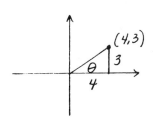

$\sin\theta = \dfrac{3}{5}$, $\cos\theta = \dfrac{4}{5}$

$\csc\theta = \dfrac{5}{3}$, $\sec\theta = \dfrac{5}{4}$, $\cot\theta = \dfrac{4}{3}$

21. $\tan \theta = \frac{3}{4}$, Q III

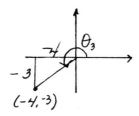

$\sin \theta = \frac{-3}{5}$, $\cos \theta = \frac{-4}{5}$

$\csc \theta = \frac{-5}{3}$, $\sec \theta = \frac{-5}{4}$, $\cot \theta = \frac{4}{3}$

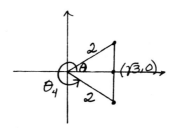

23. $\cos \theta = \frac{\sqrt{3}}{2}$

$y = \sqrt{2^2 - (\sqrt{3})^2} = 1$

Q I: $\sin \theta = \frac{1}{2}$, $\csc \theta = 2$

$\tan \theta = \frac{1}{\sqrt{3}}$, $\cot \theta = \sqrt{3}$, $\sec \theta = \frac{2}{\sqrt{3}}$

Q IV: $\sin \theta = \frac{-1}{2}$, $\csc \theta = -2$

$\tan \theta = \frac{-1}{\sqrt{3}}$, $\cot \theta = -\sqrt{3}$, $\sec \theta = \frac{2}{\sqrt{3}}$

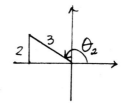

25. $\sin \theta = \frac{2}{3}$, Q II

$x = \sqrt{3^2 - 2^2} = \sqrt{5}$

$\cos \theta = -\frac{\sqrt{5}}{3}$, $\sec \theta = \frac{-3}{\sqrt{5}}$

$\tan \theta = -\frac{2}{\sqrt{5}}$, $\cot \theta = -\frac{\sqrt{5}}{2}$, $\csc \theta = \frac{3}{2}$

27. $\sin \theta = = -\frac{1}{2}$

$x = \sqrt{2^2 - 1^2} = \sqrt{3}$

Q III: $\cos \theta = \frac{-\sqrt{3}}{2}$, $\sec \theta = \frac{-2}{\sqrt{3}}$

$\tan \theta = \frac{1}{\sqrt{3}}$, $\cot \theta = \sqrt{3}$, $\csc \theta = = -2$

Q IV: $\cos \theta = \frac{\sqrt{3}}{2}$, $\sec \theta = \frac{2}{\sqrt{3}}$

$\tan \theta = -\frac{1}{\sqrt{3}}$, $\cot \theta = -\sqrt{3}$, $\csc \theta = -2$

29. $\tan \theta = \frac{1}{10}$, Q I

$r = \sqrt{10^2 + 1} = \sqrt{101}$

$\sin \theta = \frac{10}{\sqrt{101}}$, $\csc \theta = \frac{\sqrt{101}}{10}$

$\cos \theta = \frac{1}{\sqrt{101}}$, $\sec \theta = \sqrt{101}$, $\cot \theta = \frac{1}{10}$

31. $\cos \theta = \frac{12}{13}$,

$y = \sqrt{13^2 - 12^2} = 5$

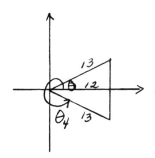

Q I: $\sin \theta = \frac{5}{13}$, $\csc \theta = \frac{13}{5}$

$\tan \theta = \frac{5}{12}$, $\cot \theta = \frac{12}{5}$, $\sec \theta = \frac{13}{12}$

Q IV: $\sin \theta = \frac{-5}{13}$, $\csc \theta = \frac{-13}{5}$

$\tan \theta = \frac{-5}{12}$, $\cot \theta = \frac{-12}{5}$, $\sec \theta = \frac{13}{12}$

33. $\tan \theta = -\pi$

$$r = \sqrt{(-\pi)^2 + 1} = \sqrt{\pi^2 + 1}$$

Q II: $\sin \theta = \dfrac{\pi}{\sqrt{\pi^2 + 1}}$, $\csc \theta = \dfrac{\sqrt{\pi^2 + 1}}{\pi}$

$\cos \theta = \dfrac{-1}{\sqrt{\pi^2 + 1}}$, $\sec \theta = -\sqrt{\pi^2 + 1}$, $\cot \theta = \dfrac{-1}{\pi}$

Q IV: $\sin \theta = \dfrac{-\pi}{\sqrt{\pi^2 + 1}}$, $\csc \theta = \dfrac{-\sqrt{\pi^2 + 1}}{\pi}$

$\cos \theta = \dfrac{1}{\sqrt{\pi^2 + 1}}$, $\sec \theta = \sqrt{\pi^2 + 1}$, $\cot \theta = -\dfrac{1}{\pi}$

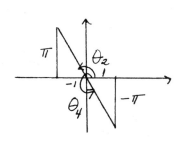

Section 6.3   Fundamental Relations

1. $\sin \theta = \frac{1}{2}$

$\csc \theta = \dfrac{1}{\sin \theta} = 2$

3. $\sec \beta = 3$

$\cos \beta = \dfrac{1}{\sec \beta} = \frac{1}{3}$

5. $\sin A = \dfrac{\sqrt{3}}{2}$

$\cos A = \sqrt{1 - \sin^2 A} = \sqrt{1 - \frac{3}{4}} = \sqrt{\frac{1}{4}} = \frac{1}{2}$

7. $\cot \theta = \sqrt{2}$

$\tan \theta = \dfrac{1}{\cot \theta} = \dfrac{1}{\sqrt{2}}$

9. $\csc \alpha = 2$

$\sin \alpha = \dfrac{1}{\csc \alpha} = \frac{1}{2}$

$\cos \alpha = \sqrt{1 - \sin^2 \alpha} = \sqrt{1 - \frac{1}{4}} = \dfrac{\sqrt{3}}{2}$

11. $\sin \phi = \frac{5}{13}, \cos \phi = \frac{12}{13}$

$$\tan \phi = \frac{\sin \phi}{\cos \phi} = \frac{\frac{5}{13}}{\frac{12}{13}} = \frac{5}{12}$$

13. $\tan \theta = \frac{1}{2}, \cos \theta = \frac{2}{15}$

$$\tan \theta = \frac{\sin \theta}{\cos \theta}$$

$$\sin \theta = \cos \theta \tan \theta = \frac{2}{\sqrt{5}} \cdot \frac{1}{2} = \frac{1}{\sqrt{5}}$$

15. $\sin x = \frac{1}{\sqrt{10}}, \cos x = \frac{3}{\sqrt{10}}$

$$\tan x = \frac{\sin x}{\cos x} = \frac{\frac{1}{\sqrt{10}}}{\frac{3}{\sqrt{10}}} = \frac{1}{3}$$

17. $\csc \beta = \frac{\sqrt{5}}{2}$

$$\sin \beta = \frac{1}{\csc \beta} = \frac{2}{\sqrt{5}}$$

$$\cos \beta = \sqrt{1-\sin^2 \beta} = \sqrt{1-\frac{4}{5}} = \frac{1}{\sqrt{5}}$$

$$\tan \beta = \frac{\frac{2}{\sqrt{5}}}{\frac{1}{\sqrt{5}}} = 2$$

19. $\cos \phi = \frac{\sqrt{2}}{3}$

$$\sin \phi = \sqrt{1-\cos^2 \phi} = \sqrt{1-\frac{2}{9}} = \frac{\sqrt{7}}{3}$$

$$\cot \phi = \frac{\cot \phi}{\sin \phi} = \frac{\frac{\sqrt{2}}{3}}{\frac{\sqrt{7}}{3}} = \sqrt{\frac{2}{7}}$$

21. $\csc \theta = 3$

$$\sin \theta = \frac{1}{\csc \theta} = \frac{1}{3}$$

$$\cos \theta = \sqrt{1-\sin^2 \theta} = \sqrt{1-\frac{1}{9}} = \frac{2\sqrt{2}}{3}, \sec \theta = \frac{3}{2\sqrt{2}}$$

$$\tan \theta = \frac{\frac{1}{3}}{\frac{2\sqrt{2}}{3}} = \frac{1}{2\sqrt{2}}, \quad \cot \theta = 2\sqrt{2}$$

23. $\cos \theta = \frac{2}{\sqrt{5}}$, $\sec \theta = \frac{\sqrt{5}}{2}$

$\sin \theta = \sqrt{1-\frac{4}{5}} = \frac{1}{\sqrt{5}}$, $\csc \theta = \sqrt{5}$

$\tan \theta = \dfrac{\frac{1}{\sqrt{5}}}{\frac{2}{\sqrt{5}}} = \frac{1}{2}$, $\cot \theta = 2$

25. $\tan \beta = \sqrt{2}$, $\cot \beta = \frac{1}{\sqrt{2}}$

$\sin^2\beta + \cos^2\beta = 1$

$\tan^2\beta + 1 = \frac{1}{\cos^2\beta} = \sec^2\beta$

$\sec \beta = \sqrt{2+1} = \sqrt{3}$, $\cos \beta = \frac{1}{\sqrt{3}}$

$\sin \beta = \sqrt{1-\frac{1}{3}} = \sqrt{\frac{2}{3}}$, $\csc \beta = \sqrt{\frac{3}{2}}$

27. $\cos x = 0.1155$

$\sec x = \frac{1}{.1155} = 8.658$

29. $\csc A = 1.902$

$\sin A = \frac{1}{1.902} = 0.5258$

1.

3.

$$x = \sqrt{2^2-1} = \sqrt{3}$$

$\sin 30° = \frac{1}{2}$, $\csc 30° = 2$

$\cos 30° = \frac{\sqrt{3}}{2}$, $\sec 30° = \frac{2}{\sqrt{3}}$

$\tan 30° = \frac{1}{\sqrt{3}}$, $\cot 30° = \sqrt{3}$

$r = 1$

$\sin 0° = \frac{0}{1} = 0$, $\csc 0° =$ undefined

$\cos 0° = \frac{1}{1} = 1$, $\sec 0° = 1$

$\tan 0° = 0$, $\cot 0° =$ undefined

5. $\sin 20° = \cos (90°-20°) = \cos 70°$

7. $\tan 20°15' = \cot (90°-20°15') = \cot 69°45'$

9. $\sec 81° = \csc (90°-81°) = \csc 9°$

11. $\cos 17.5° = \sin (90°-17.5°) = \sin (72.5°)$

13. $\tan 60.4° = \cot (90°-60.4°) = \cot 29.6°$

15. $2 \sin 30° + 5 \cos 45° - 2 = 0$

$2 \left(\frac{1}{2}\right) + 5 \left(\frac{\sqrt{2}}{2}\right) - 2 = \frac{5\sqrt{2}-2}{2}$

17. $\tan 45° \sin 270° = (1)(-1) = -1$

19. $\csc^2 60° + 3 \cot^2 270° = \left(\frac{2}{\sqrt{3}}\right)^2 + 3(0^2) = \frac{4}{3}$

21. $\frac{\sec^2 180°}{\sin 180°} = \frac{(-1)^2}{0} =$ undefined

23. $(2 \sin 60° + \sqrt{3} \cos 180°)^2 = \left(2\left(\frac{\sqrt{3}}{2}\right) + \sqrt{3}\,(-1)\right)^2 = 0$

37. $x = 15 \tan 30° = 15\left(\frac{1}{\sqrt{3}}\right) = 5\sqrt{3}$

Section 6.5   Reference Angles

1. $\alpha = 180° - 100° = 80°$

3. $\alpha = 360° - 300° = 60°$

5. $\alpha = 210° - 180° = 30°$

7. $480° = 360° + 120°$

$\alpha = 180° - 120° = 60°$

9. $\alpha = 180° - 135° = 45°$

11. $\alpha = \frac{5\pi}{4} - \pi = \frac{\pi}{4}$

13. $19\frac{\pi}{6} = 2\pi + 7\frac{\pi}{6}$

$\alpha = \frac{7\pi}{6} - \pi = \frac{\pi}{6}$

15. $\cos 315° = \cos 45° = \frac{\sqrt{2}}{2}$

17. $\sin 150° = \sin 30° = \frac{1}{2}$

19. $\tan 300° = -\tan 60° = -\sqrt{3}$

21. $\csc 210° = -\csc 30° = -2$

23. $\sec \frac{7\pi}{4} = \sec \frac{\pi}{4} = \sqrt{2}$

25. $\sin \frac{4\pi}{3} = -\sin \frac{\pi}{3} = \frac{-\sqrt{3}}{2}$

27. $\cot 510° = \cot 150° = -\cot 30° = -\sqrt{3}$

29. $\cos 1050° = \cos 330° = \cos 30° = \frac{\sqrt{3}}{2}$

31. $\tan \frac{19\pi}{6} \sec = \tan \frac{7\pi}{6} = \tan \frac{\pi}{6} = \frac{1}{\sqrt{3}}$

33. $\sin \frac{21\pi}{4} = \sin \frac{5\pi}{4} = -\sin \frac{\pi}{4} = \frac{-\sqrt{2}}{2}$

27. $\tan \theta = .4414$, $\cos \theta < 0 \rightarrow \theta$ in Q III

   $\alpha = 23.8°$, $\theta = 180° + 23.8° = 203.8°$

29. $\sin \theta = -.4253$, $\tan \theta < 0 \rightarrow \theta$ in Q IV

   $\alpha = 25.2°$, $\theta = 360° - 25.2° = 334.8°$

33. $\sec \theta = 1.345$, $\tan \theta < 0 \rightarrow \theta$ in Q IV

   $\alpha = 42.0°$, $\theta = 360 - 42° = 318°$

35. $\tan \theta = -0.3378$, θ in Q II, IV

   $\alpha = -18.7°$, $\theta_2 = 161.3°$, $\theta_4 = 341.3°$

37. $\csc \theta = 1.505$, θ in Q I, II

   $\alpha = 41.6°$, $\theta_1 = 41.6°$, $\theta_2 = 138.4°$

39. $\cot \theta = -0.5137$, θ in Q II, IV

   $\alpha = 62.8°$, $\theta_2 = 117.2°$, $\theta_4 = 297.2°$

Section 6.7   Values of the Trigonometric Functions from a Table

11. $\cos 125° = -\cos 55°$ 13. $\sin 225° = -\sin 45°$

15. $\tan 1243° = \tan 163° = -\tan 17°$

17. $\sec 10 = \sec (10-2\pi) = -\sec (10-3\pi) = -\sec .58$

19. $\cos 45 = \cos (45-14\pi) = \cos 1.04$

21. $\sin 154° = \sin 26° = .4384$ 23. $\tan 96° = -\tan 84° = -9.5144$

25. $\sec 163°20' = -\sec 16°40' = -1.0439$

27. $\cos 285°50' = \cos 74°10' = .2728$

29. $\sin 247°36' = -\sin 67°36'$

$$10 \left[6 \left[\begin{array}{l} \sin 67°30' \\ \sin 67°36' \end{array} \begin{array}{l} = .9239 \\ = x \end{array} \right] y \\ \quad\quad \sin 67°40' \quad = .9250 \right] .0011$$

$$\frac{6}{16} = \frac{y}{.0011}, \; y = .00066$$

$$\sin 247°36' = -(.9239 + .00066) = -.9246$$

31. $\sin 100 = \sin 5.76 = -\sin .531$

$$.01 \left[.001 \left[\begin{array}{l} \sin .53 \\ \sin .531 \end{array} \begin{array}{l} = .5055 \\ = x \end{array} \right] y \\ \quad\quad \sin .54 \quad = .5141 \right] .0086$$

$$\frac{.001}{.01} = \frac{y}{.0086}, \; y = .00086$$

$$\sin 100 = -(.5055 + .0009) = -.5064$$

33. $\sin 4.22 = -\sin 1.078$

$$.01 \left[.008 \left[\begin{array}{l} \sin 1.07 \\ \sin 1.078 \end{array} \begin{array}{l} = .8772 \\ = x \end{array} \right] y \\ \quad\quad \sin 1.08 \quad = .8820 \right] .0048$$

$$\frac{.008}{.01} = \frac{y}{.0048}, \; y = .004$$

$$\sin 1.078 = .8772 + .004 = .8812$$
$$\sin 4.22 = -.8812$$

35. $\csc 20 = \csc 1.15 = 1.096$

37.

$$10 \left[.05 \left[\begin{array}{l} \cot 1.33 \\ \cot 1.335 \end{array} \begin{array}{l} = .2456 \\ = x \end{array} \right] y \\ \quad\quad \cot 1.34 \quad = .2350 \right] .0106$$

$$\frac{.05}{10} = \frac{y}{.0106}, \; y = .0053$$

$$\cot 1.335 = .2456 - .0053 = .2403$$

39. $\cos 3.01 = -\cos .13 = -.9916$

41. $\sin \theta = -.6157,\ \cos \theta > 0,\ \rightarrow \theta$ in Q IV

$$.01 \left[x \left[\begin{array}{lll} \sin .66 & = & .6131 \\ \sin \alpha & = & .6157 \end{array} \right] .0026 \atop \begin{array}{lll} \sin .67 & = & .6210 \end{array} \right] .0079$$

$$\frac{x}{.01} = \frac{.0026}{.0079},\ x = .0033$$

$$\alpha = .66 + .0033 = .6633$$
$$\theta = 2\pi - .6633 = 5.620$$

43. $\cot \theta = .9004,\ \sin \theta < 0 \rightarrow \theta$ in Q III

$$.01 \left[x \left[\begin{array}{lll} \cot .84 & = & .8964 \\ \cot \alpha & = & .9004 \end{array} \right] .004 \atop \begin{array}{lll} \cot .83 & = & .9146 \end{array} \right] .0182$$

$$\frac{x}{.01} = \frac{.004}{.0182},\ x = .0022$$

$$\alpha = .84 - .0022 = .8378$$
$$\theta = \pi + .8378 = 3.979$$

45. $\tan \theta = -1.804,\ \sec \theta > 0 \rightarrow \theta$ in Q IV

$$.01 \left[x \left[\begin{array}{lll} \tan 1.06 & = & 1.784 \\ \tan \alpha & = & 1.804 \end{array} \right] .002 \atop \begin{array}{lll} \tan 1.07 & = & 1.827 \end{array} \right] .0043$$

$$\frac{x}{.01} = \frac{.002}{.0043},\ x = .0047$$

$$\alpha = 1.0647,\ \theta = 2\pi - 1.065 = 5.219$$

47. $\sin \theta = -.4253,\ \tan \theta < 0 \rightarrow \theta$ in Q IV
$$\alpha = 25°10',\ \theta = 360° - 25°10' = 334°50'$$

49. $\cot \theta = 3.0326,\ \csc \theta < 0 \rightarrow \theta$ in Q III

$$10' \left[x \left[\begin{array}{lll} \cot 18°10' & = & 3.0475 \\ \cot \alpha & = & 3.0326 \end{array} \right] -.0149 \atop \begin{array}{lll} \cot 18°20' & = & 3.0178 \end{array} \right] -.0297$$

$$\frac{x}{10} = \frac{-.0149}{-.0297},\ x = 5.02$$

$$\alpha = 18°10' + 5' = 18°15'$$
$$\theta = 180° + 18°15' = 198°15'$$

51. $\tan \theta = -6.8269$, $\csc \theta > 0 \rightarrow \theta$ in Q II

$\alpha = 81°40'$, $\theta = 180° - 81°40' = 98°20'$

53. $\sec \theta = -1.1122$, $\tan \theta < 0$, θ in Q II

$$10' \left[x \left[\begin{array}{lll} \sec 25°50' & = & 1.1110 \\ \sec \alpha & = & 1.1122 \end{array} \right].0012 \atop \sec 26° \quad = \quad 1.1126 \right].0016$$

$$\frac{x}{10} = \frac{.0012}{.0016}, \ x = 7.5$$

$$\alpha = 25°58', \ \theta = 180° - 25°58' = 154°2'$$

55. $\cot \theta = .5454$, $\sin \theta < 0 \rightarrow \theta$ in Q III

$$10' \left[x \left[\begin{array}{lll} \cot 61°20' & = & .5467 \\ \cot \alpha & = & .5454 \end{array} \right].0013 \atop \cot 61°30' \quad = \quad .5430 \right].0037$$

$$\frac{x}{10} = \frac{.0013}{.0037}, \ x = 3.51$$

$$\alpha = 61°24', \ \theta = 180° + 61°24' = 241°24'$$

Chapter 6   Review

7. $36° = 36 \cdot \frac{\pi}{180} = \frac{\pi}{5}$

9. $110° = 110\frac{\pi}{180} = \frac{11\pi}{18}$

11. $\frac{5\pi}{12} = \frac{5\pi}{12} \cdot \frac{180}{\pi} = 75°$

13. $\frac{7\pi}{4} = \frac{7\pi}{4} \cdot \frac{180}{\pi} = 315°$

15. $\sin \theta = \frac{5}{13}$, $\cos \theta > 0$

$x = \sqrt{13^2 - 5^2} = 12$

$\cos \theta = \frac{12}{13}$, $\sec \theta = \frac{13}{12}$

$\tan \theta = \frac{5}{12}$, $\cot \theta = \frac{12}{5}$, $\csc \theta = \frac{13}{5}$

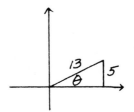

17. $r = \sqrt{5^2 + 2^2} = \sqrt{29}$

$\sin \theta = \frac{5}{\sqrt{29}}$, $\cos \theta = \frac{-2}{\sqrt{29}}$, $\tan \theta = \frac{-5}{2}$

$\csc \theta = \frac{\sqrt{29}}{5}$, $\sec \theta = \frac{-\sqrt{29}}{2}$, $\cot \theta = \frac{-2}{5}$

(-2,5)

19. $\cos \theta = \frac{-4}{5}$, $\tan \theta > 0 \rightarrow \theta$ in Q III

$y = \sqrt{5^2 - 4^2} = 3$

$\sin \theta = \frac{-3}{5}$, $\csc \theta = \frac{-5}{3}$

$\tan \theta = \frac{3}{4}$, $\cot \theta = \frac{4}{3}$, $\sec \theta = \frac{-5}{4}$

21. $\alpha = 360° - 315°18' = 44°42'$

23. $\alpha = 241.9° - 180° = 61.9°$

25. $\alpha = 2\pi - 4.75 = 1.53$

35. $\sin x = -.8102$, x in Q III and Q IV

$\alpha = 54°7'$

$x = 360° - 54°7' = 305°53'$

$x = 180° + 54°7' = 234°7'$

37. $\sec \phi = 2.603$, $\sin \phi < 0 \rightarrow \phi$ in Q IV

$\alpha = 67.4°$, $\phi = 360° - 67.4° = 292.6°$

39. $\tan \beta = -.7761$, $\cos \beta > 0 \rightarrow \beta$ in Q IV

$\alpha = .6600$, $\beta = 2\pi - .6600 = 5.62$

Chapter 7 The Solution of Triangles

Section 7.1 Solution of Right Triangles

1. $c = \sqrt{42.6^2 + 70.5^2} = 82.4$

   $\tan A = \frac{42.6}{70.5} = .6043$

   $A = 31.1°$, $B = 90° - 31.1° = 58.9°$

3. $b = \sqrt{20^2 - 12^2} = 16.0$

   $\sin A = \frac{12}{20} = 1.2$

   $A = 36.9°$, $B = 90° - 36.9° = 53.1°$

5. $B = 90° - 9°20' = 80°40'$

   $\tan 9°20' = \frac{5.00}{b}$, $b = \frac{5}{\tan 9°20'} = 30.4$

   $c = \sqrt{5.00^2 + 30.4^2} = 30.8$

7. $\tan 23.5° = \frac{x}{3.15}$

   $x = 3.15 \tan 23.5° = 13.7$

9. $\tan A = \frac{6}{10} = .6$

   $A = 31°$

   $90° = 6$ hours

   $1° = \frac{6}{90}$ hours

   $31° = 31\left(\frac{6}{90}\right) = 2$ hours

   Time is 8 a.m.

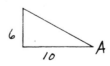

11. $\theta = 180° - 127° = 53°$

$\dfrac{h}{6.4} = \sin 53°$

$h = 6.4 \sin 53° = 5.1$

13. $\tan B = \dfrac{7}{11} = .6364$

$B = 32.5°, \; A = 90° - 32.5° = 57.5°$

15. $\cos P = \dfrac{190}{500} = .3800,$

$P = 67.7°, \; S = 90° - 67.7° = 22.3°$

17. $\angle BAD = 33.6°, \; \angle CAD = 32.1°$

$BD = x + h$

$\tan 32.1° = \dfrac{x}{550}, \; x = 500 \tan 32.1° = 345.0$

$\tan 33.6° = \dfrac{x + h}{550}, \; x + h = 550 \tan 33.6°$

$h = 550 \tan 33.6° - 345 = 20.4$

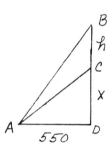

19. $\tan D = \dfrac{220}{BD}, \; BD = \dfrac{220}{\tan 28°15'} = 409.4$

$BC = \dfrac{220}{\tan C} = \dfrac{220}{\tan 36°20'} = 299.1$

$CD = BD - BC = 110.3$

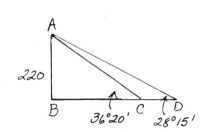

21. $\tan 49.1° = \frac{h}{.32}$

   $h = .32 \tan 49.1° = .369$

   $.369$ cm $\rightarrow .369(2500)$ m $= 923.5$ m

23. $\sin 8.7° = \frac{5800}{x}$

   $x = \frac{5800}{\sin 8.7°} = 38{,}344$

   $t = \frac{38{,}344}{265} = 145$ sec $= 2$ min 5 sec

25. $w = 2x + 4.2$

   $\sin 51.5° = \frac{x}{28.3}$

   $x = 28.3 \sin 51.5° = 22.15$

   $w = 44.3 + 4.2 = 48.5$

27. $\tan 30° = \frac{x}{6}$

   $x = 6 \tan 30° = 3.46$

   $y = 2(3.46) = 6.93$

   $c = y + 2x = 6.93 + 2(3.46) = 13.8$

29. $\sin 22.5° = \frac{x}{1.5 + x}$

   $x = (1.5 + x)(.3827)$

   $x = .93$

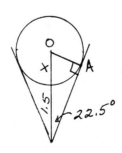

31. $\cos 22.6° = \frac{r}{100+r}$

$(100+r)(.9232) = r$

$r = 1202 \simeq 1200$

Section 7.2   The Law of Cosines

1. $C^2 = 67^2+45^2-2(67)(45) \cos 35° = 1574.5$

$C = 39.7$

3. $C^2 = 10.5^2+40.8^2-2(40.8)(10.5) \cos 120° = 2203.29$

$C = 46.9$

5. $a^2 = 42.0^2+38.0^2-2(42)(38) \cos 135.3° = 5476.87$

$a = 74.0$

7. $\cos C = \dfrac{7.23^2+6^2-8.61^2}{2(7.23)(6)} = .1629,\ C = 80.6°$

9. $\cos A = \dfrac{10^2+14^2-18^2}{2(10)(14)} = -.1,\ A = 95.7°$

11. $\cos B = \dfrac{170^2+120^2-250^2}{2(170)(120)} = -.4706,\ B = 118.1°$

13. $c^2 = 4.21^2+1.84^2-2(4.21)(1.84) \cos 30.7° = 7.788$

$c = 2.79$

$\cos A = \dfrac{1.84^2+2.79^2-4.21^2}{2(1.84)(2.79)} = -.6384,\ A = 129.7$

$B = 180°-30.7°-129.7° = 19.6°$

15. $c^2 = 120^2+145^2-2(120)(145) \cos 94°25' = 38106.93$

$c = 195.2$

$\cos A = \dfrac{145^2+195.2^2-120^2}{2(145)(195.2)} = .7901,\ A = 37°48'$

$B = 180°-37°48'-94°25' = 47°47'$

17. $\cos A = \dfrac{4^2+3^2-2^2}{2(4)(3)} = .875,\ A = 29.0$

$\cos B = \dfrac{4^2+2^2-3^2}{2(4)(2)} = .6875,\ B = 46.6°$

$C = 180°-29°-46.6° = 104.4°$

19. $\angle\ B = 180°-25.4° = 154.6°$

$\overline{AC}^2 = 2.5^2+1.4^2-2(2.5)(1.4) \cos 154.6° = 14.53$

$\overline{AC} = 3.81$

21. $\cos A = \dfrac{40^2+80^2-50^2}{2(40)(80)} = .8594,\ A = 30.8°$

23. $\cos A = \dfrac{5.5^2+5.5^2-1.3^2}{2(5.5)(5.5)} = .9721,\ A = 13.6$

25. $\overline{AC}^2 = 3500^2+4000^2-2(3500)(4000) \cos 60°$

$AC = 3755$

$\cos A = \dfrac{3500^2+3775^2-4000^2}{2(3500)(3775)} = .3973$

$A = 66.6°,\ C = 180°-60°-66.6° = 53.4°$

27. $\overline{AB}^{\,2} = 200^2 + 400^2 - 2(200)(400)\cos 60°$

$AB = 346.4$

Section 7.3 The Law of Sines

1. $\dfrac{b}{\sin 48°} = \dfrac{10}{\sin 32°}$

$b = 14$

$C = 180° - 48° - 32° = 100°$

$\dfrac{c}{\sin 100°} = \dfrac{10}{\sin 32°}$

$c = 18.6$

3. $\dfrac{\sin \beta}{5.15} = \dfrac{\sin 45.2°}{8.82}$

$B = 24.5°, \ \cancel{155.5°}$ (too large)

$C = 180° - 24.5° - 45.2° = 110.3°$

$\dfrac{c}{\sin 110.3°} = \dfrac{8.82}{\sin 45.2°}$

$c = 11.7$

5. $\dfrac{\sin B}{12.2} = \dfrac{\sin 35.6}{17.5}$

  $B = 23.9°, \cancel{156.°}$ (too large)

  $C = 180° - 23.9° - 35.6° = 120.5°$

  $\dfrac{c}{\sin 120.5°} = \dfrac{17.5}{\sin 35.6°}$

  $c = 25.9$

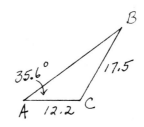

7. $\dfrac{\sin B}{5.09} = \dfrac{\sin 120°50'}{6.61}$

  $B = 41°20', \cancel{138°40'}$ (too large)

  $C = 180° - 120°50' - 41°20' = 17°50'$

  $\dfrac{c}{\sin 17°50'} = \dfrac{6.61}{\sin 120°50'}$

  $c = 2.35$

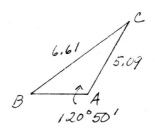

9. $\dfrac{\sin B}{18.3} = \dfrac{\sin 53°}{30.2}$

  $B = 28.9°, \cancel{151.1°}$ (too large)

  $A = 180° - 28.9° - 53° = 98.1°$

  $\dfrac{a}{\sin 98.1°} = \dfrac{30.2}{\sin 53°}$

  $a = 37.4$

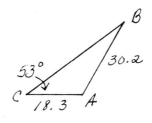

11. $\dfrac{\sin A}{25.0} = \dfrac{\sin 122°}{30.0}$

  $A = 45.0°, \cancel{135°}$ (too large)

  $C = 180° - 45° - 122° = 13°$

  $\dfrac{c}{\sin 13°} = \dfrac{30}{\sin 122°}$

  $c = 7.98$

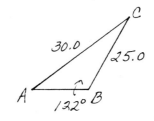

13. $\dfrac{c}{\sin 110°} = \dfrac{40}{\sin 50°}$

  $c = 49.1$

  $A = 180° - 110° - 50° = 20°$

  $\dfrac{a}{\sin 20°} = \dfrac{40}{\sin 50°}$

  $a = 17.9$

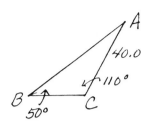

15. $C = 180° - 43.4° - 48° = 88.6°$

  $\dfrac{b}{\sin 48°} = \dfrac{61.3}{\sin 88.6°}$, $b = 45.6$

  $\dfrac{a}{\sin 43.4°} = \dfrac{61.3}{\sin 88.6°}$, $a = 42.1$

17. $\dfrac{\sin B}{6.00} = \dfrac{\sin 26°20'}{2.00}$

  $\sin B = \dfrac{6 \sin 26°20'}{2.00} = 1.3308$

  No solution

19. $\dfrac{\sin C}{8} = \dfrac{\sin 150°}{15}$

  $C = 15.5°, \cancel{164.5°}$ (too large)

  $B = 180° - 150° - 15.5° = 14.5°$

  $\dfrac{b}{\sin 14.5°} = \dfrac{15}{\sin 150°}$, $b = 7.53$

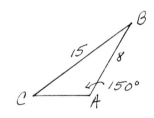

21. $\dfrac{\sin C}{74°} = \dfrac{\sin 140°}{60}$

  $C = 52.4°, 127.6°$ — Both too large

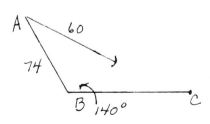

23. $\dfrac{\sin A}{20} = \dfrac{\sin 28°}{15}$

    $A_1 = 38.8°,\ 141.2° = A_2$

    $B_1 = 180° - 38.8° - 28° = 113.3°$

    $\dfrac{\overline{CA}}{\sin 113.3°} = \dfrac{15}{\sin 28°}$

    $\overline{CA} = 29.4$

  or

    $B_2 = 180° - 141.2° - 28° = 10.8°$

    $\dfrac{\overline{CA_1}}{\sin 10.8°} = \dfrac{15}{\sin 28°}$

    $\overline{CA_1} = 5.96$

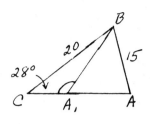

25. $\dfrac{\sin B}{400.5} = \dfrac{\sin 30.04°}{325.1}$

    $B = 38.08°,\ B_1 = 141.92°$

    $B = 180° - 30.04° - 38°08' = 111.88°$

    $\dfrac{\overline{AB}}{\sin 111.92°} = \dfrac{325.1}{\sin 30°},\ \overline{AB} = 602.6$

  or

    $C_1 = 180° - 30.04° - 141.92° = 8.04°$

    $\dfrac{\overline{AB_1}}{\sin 8.08°} = \dfrac{325.1}{\sin 30°},\ \overline{AB_1} = 90.82$

27. $B = 70°,\ A = 180° - 2(70°) = 40°$

    $\dfrac{a}{\sin 40°} = \dfrac{100}{\sin 70°},\ a = 68.4$

29. $\dfrac{\sin B}{.700} = \dfrac{\sin 72°10'}{.900}$

   $B = 47°50', \cancel{132°10'}$ (too large)

   $C = 180° - 72°10' - 47°50' = 60°$

   $\dfrac{c}{\sin 60°} = \dfrac{.900}{\sin 72°10'}$

   $C = .818$

31. $\theta = 180° - 42.7° = 137.3°$

   $\dfrac{\sin A}{10.8} = \dfrac{\sin 137.3°}{15.5}$

   $A = 28.2°, \cancel{151.8°}$ (too large)

   $B = 180° - 28.2° - 137.3° = 14.5°$

   $\dfrac{b}{\sin 14.5°} = \dfrac{15.5}{\sin 137.3°}$

   $b = 5.72$

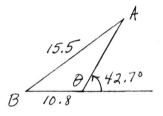

33. $B = 180° - 124.6° = 55.4°$

   $a = 180° - 52.1° - 55.4° = 72.5°$

   $\dfrac{a}{\sin 52.1°} = \dfrac{230}{\sin 72.5°}$, $a = 190.3$

   $t = \dfrac{190.3}{125} = 1.52$ hours $= 1$ hour 31 minutes

35. $\alpha = 90° - 41° = 49°$

   $\beta = 90° - 75° = 15°$

   $\gamma = 180° - 49° - 15° = 116°$

   $\dfrac{a}{\sin 15°} = \dfrac{25}{\sin 116°}$

   $a = 7.20$

   $x = 7.2 \sin 41° = 4.72$

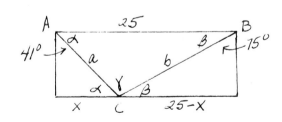

37. $\beta = 75.4° - 30° = 45.4°$

$\alpha = 43.5° - 30° = 13.5°$

$\phi = 180° - 60° = 120°$

$\varphi = 180° - 120° - 45.4° = 14.6°$

$\theta = 90° - 43.5° - 14.6° = 31.9°$

$\dfrac{h}{\sin \alpha} = \dfrac{1500}{\sin \theta}$

$h = \dfrac{1500 \sin 13.5°}{\sin 31.9°} = 662.6$

$\dfrac{x}{\sin \beta} = \dfrac{662.6}{\sin \phi}$

$x = \dfrac{662.6 \sin 45.4°}{\sin 120°} = 545 \text{ (antenna)}$

$\dfrac{k}{\sin \varphi} = \dfrac{662.6}{\sin 120°}$

$k = \dfrac{662.6 \sin 14.6°}{\sin 120°} = 193$

$\sin 30° = \dfrac{y}{1500 + 193}$

$y = 847 \text{ (hill)}$

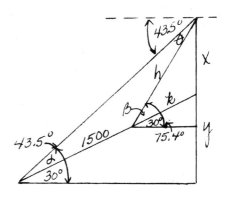

Chapter 7   Review

1. $B = 90° - 32° = 58°$

$\tan 32° = \dfrac{3.0}{b}$

$b = 4.8$

$C = \sqrt{3^2 + 4.8^2} = 5.7$

3. $\sin A = \dfrac{29}{41}$

$B = 90° - 45° = 45°$

$b = 29$

5. $x = 5 - 3.25 = 1.75$

$y = 12.50 - 8.20 = 4.30$

$\tan \theta = \dfrac{4.30}{1.75}, \ \theta = 67.9°$

7. $\sin 5°25' = \frac{x}{5000}$

   $x = 472$

9. $\sin 23° = \frac{y}{15}$, $y = 5.86$

   $\cos 23° = \frac{x}{15}$, $x = 13.8$

11. $a^2 = 17^2 + 28^2 - 2(17)(28) \cos 29°$

    $a = 15.5$

    $\cos B = \frac{15.5^2 + 28^2 - 17^2}{2(15.5)(28)}$

    $B = 32.1°$

    $C = 180° - 32.1° - 29° = 118.9°$

13. $\cos A = \frac{7.9^2 + 15.4^2 - 11.2^2}{2(7.9)(15.4)}$

    $A - 44.3$

    $\cos B = \frac{11.2^2 + 15.4^2 - 7.9^2}{2(11.2)(15.4)}$

    $B = 29.5°$

    $C = 180° - 44.3° - 29.5° = 106.2°$

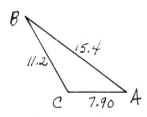

15. $\cos A = \dfrac{5.88^2 + 17.8^2 - 23^2}{2(5.88)(17.8)}$

$A = 148°$

$\cos B = \dfrac{23^2 + 17.8^2 - 5.88^2}{2(23)(17.8)}$

$B = 7.8°$

$C = 180° - 148° - 7.8° = 24.2°$

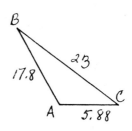

17. $\dfrac{c}{\sin 27.6°} = \dfrac{3120}{\sin 112.2°}$

$C = 1561$

$B = 180° - 112.2° - 27.6° = 40.2°$

$\dfrac{b}{\sin 40.2°} = \dfrac{3120}{\sin 112.2°},\ b = 2175$

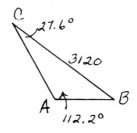

19. $\dfrac{\sin B}{12} = \dfrac{\sin 25°}{15}$

$B - 19.8°,\ \cancel{160.2°}\ \text{(too large)}$

$C = 180° - 25° - 19.8° = 135.2°$

$\dfrac{c}{\sin 135.2°} = \dfrac{15}{\sin 25°}$

$C = 25.0$

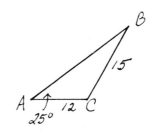

21. $\dfrac{\sin A}{23} = \dfrac{\sin 63°50'}{12}$

$\sin A = 1.7202,\ \text{no solution}$

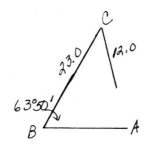

23. $\alpha = 180° - 45° = 135°$

$\beta = 180° - 30° - 135° = 15°$

$\dfrac{x}{\sin 30°} = \dfrac{1}{\sin 15°}$

$x = 1.93$

25. $\alpha = 180° - 58° = 122°$

$\beta = 180° - 122° - 30° = 28°$

$\dfrac{x}{\sin 30°} = \dfrac{100}{\sin 28°}$, $x = 106.5$

$\sin 58° = \dfrac{h}{106.5}$, $h = 90.312$

Chapter 8   Analytic Trigonometry

<u>Section 8.2   More on the Sine and Cosine Functions</u>

5. $\text{per} = \dfrac{2\pi}{\frac{2}{3}} = 3\pi$

7. $\text{per} = \dfrac{2\pi}{\pi} = 2$

9. $\text{p.s.} = \dfrac{-\frac{\pi}{3}}{1} = -\dfrac{\pi}{3}$

11. $\text{per} = \dfrac{2\pi}{\frac{1}{2}} = 4\pi$

$\text{p.s.} = \dfrac{\frac{\pi}{2}}{\frac{1}{2}} = \pi$

13. $\text{per} = \dfrac{2\pi}{2} = \pi$

$\text{p.s.} = -\dfrac{\pi}{2}$

15. $\text{per} = \dfrac{2\pi}{\frac{1}{3}} = 6\pi$

17. $\text{per} = \dfrac{2\pi}{\pi} = 2$

$\text{p.s.} = \dfrac{\frac{\pi}{4}}{\pi} = \dfrac{1}{4}$

$\text{p.s.} = \dfrac{-\frac{\pi}{3}}{\frac{1}{3}} = -\pi$

19. $\text{per} = \dfrac{2\pi}{4} = \dfrac{\pi}{2}$

21. $-\sin x = \sin(-x)$

$\text{p.s.} = \dfrac{\pi}{4}$

$-\sin(x+1) = \sin\left[-(x+1)\right]$

$\sin(-x) = \sin(x+\pi)$

$\sin\left[-(x+1)\right] = \sin\left[x+1+\pi\right]$

$\text{per} = 2\pi, \text{amp} = 1$

$\text{p.s.} = -(1+\pi)$

23. $-\sin x = \sin(-x)$

$-\sin\left(2\pi x + \frac{1}{2}\right) = \sin\left[-\left(2x\pi + \frac{1}{2}\right)\right]$

$\sin(-x) = \sin(x + \pi)$

$\sin\left(-\left(2\pi x + \frac{1}{2}\right)\right) = \sin\left(2\pi x + \frac{1}{2} + \pi\right)$

$\text{per} = \frac{2\pi}{2\pi} = 1$, amp $= 1$, p.s. $= \dfrac{-\frac{1}{2} - \pi}{2\pi} = -.58$

25. $\cos x = \sin\left(\frac{\pi}{2} - x\right)$

$-\cos(\pi x + 1) = -\sin\left(\frac{\pi}{2} - (\pi x + 1)\right) = -\sin\left[-\left(\pi x + 1 - \frac{\pi}{2}\right)\right]$

But $-\sin x = \sin(-x)$

$-\cos(\pi x + 1) = \sin\left(\pi x + 1 - \frac{\pi}{2}\right)$

$\text{per} = \frac{2\pi}{\pi} = 2$, amp $= 1$, p.s. $= \dfrac{-\left(1 - \frac{\pi}{2}\right)}{\pi} = +.18$

27. $y = A \cos(Bx + c)$

$\text{per} = \frac{5\pi}{6} + \frac{\pi}{6} = \pi$, $B = 2$

amp $= 1$, $A = 1$

p.s. $= \quad -\frac{\pi}{6} = -\frac{C}{B}$

$-\frac{\pi}{6} = -\frac{C}{2}$, $C = \frac{\pi}{3}$

$y = \cos\left(2x + \frac{\pi}{3}\right) = \cos 2\left(x + \frac{\pi}{6}\right)$

29. $y = A \sin(Bx + c)$

amp $= 2$, $A = -2$

$\text{per} = 4\pi$, $B = \frac{1}{2}$

p.s. $= 0$

$y = -2 \sin\left(\frac{1}{2}x\right)$

31. $\text{per} = \dfrac{2\pi}{\frac{\pi}{20}} = 40$

p.s. $= 0$, amp. $= 1$

Section 8.3 Sinusoidal Models

1a. $x = 15 \cos 2t$

$\text{per} = \frac{2\pi}{2} = \pi$

1b. $x = 15 \cos 2(1) = -6.24$

6.24 cm below rest position

3a. $w = \dfrac{6 \text{ rev}}{\min} \cdot \dfrac{2\pi \text{ rad}}{\text{rev}} \cdot \dfrac{1 \min}{60 \text{ sec}} = \dfrac{\pi \text{ rad}}{5 \text{ sec}}$

$r = 7, \; t_0 = 2$

$h = 6 + 7 \cos \frac{\pi}{5}(t-2)$

3b. per $= \dfrac{2\pi}{\frac{\pi}{5}} = 10$, p.s. $= 2$

3c. $h = 0 = 6 + 7 \cos \frac{\pi}{5}(t-2)$

$\cos \frac{\pi}{5}(t-2) = -\frac{6}{7}$

$\frac{\pi}{5}(t-2) = 2.6$, or $2\pi - 2.6$

2.6 goes under water

$2\pi - 2.6$ comes out of water

$\frac{\pi}{5}(t-2) = 2\pi - 2.6$

$t = (2\pi - 2.6)\frac{5}{\pi} + 2 = 7.86$

5a. $y = A \cos (Bt + C) + D$

$\frac{1}{2}$ per.$= 3$, per$= 6$, $\frac{2\pi}{B} = 6$, $B = \frac{\pi}{3}$

Amp $= \dfrac{17 + 23}{2} = 20 \qquad D = 17 - 20 = -3$

$y = -20 \cos \left(\frac{\pi}{3}t + C\right) - 3$

p.s. $= 2 = \dfrac{-C}{\frac{\pi}{3}}$, $C = \dfrac{-2\pi}{3}$

$y = -20\cos \frac{\pi}{3}(t-2) - 3$

5c. $y(2.8) = -3 - 20 \cos \frac{\pi}{3}(.8) = -16.38$

$y(6.3) = -3 - 20 \cos \frac{\pi}{3}(4.3) = 1.16$

$y(15) = -3 - 20 \cos \frac{\pi}{3}(13) = -13$

5d. $y(0) = -3-20 \cos \frac{\pi}{3}(-2) = 7$

5e. $0 = -3-20 \cos \frac{\pi}{3}(t-2)$

$\quad\quad \cos \frac{\pi}{3}(t-2) = -\frac{3}{20}$

$\quad\quad\quad \frac{\pi}{3}(t-2) = 1.72, \ t = 3.64$

7. Deer: $\quad$ per. $= \frac{2\pi}{.4} = 15.7$

$\quad\quad\quad\quad$ amp $= 400$, mean value $= 1500$

$\quad$ Pumas: $\quad$ per $= \frac{2\pi}{.4} = 15.7$

$\quad\quad\quad\quad$ amp $= 200$, mean value $= 500$

$\quad\quad\quad\quad$ p.s. $= \frac{.8}{.4} = 2$

9. $\quad$ per $= \frac{2\pi}{\frac{2\pi}{365}} = 365$, amp. $= 38$

$\quad$ mean value $= 55$

$\quad$ p.s. $= 100$ days $= 31$ days in Jan.

$\quad\quad\quad\quad\quad\quad + 28$ days in Feb.

$\quad\quad\quad\quad\quad\quad + 31$ days in Mar.

$\quad\quad\quad\quad\quad\quad + 10$ days in Apr.

11a. $\quad$ per $= \frac{1948-1750}{18} = 11$

11c. $\quad$ amp $= \frac{110-10}{2} = 50 = A$

$\quad\quad$ mean value $= \frac{10+110}{2} = 60 = D$

$\quad\quad$ per $= 11 = \frac{2\pi}{B}, \ B = \frac{2\pi}{11}$

$\quad\quad$ p.s. $= 1948$

$\quad\quad$ $y = 60+50 \cos \frac{2\pi}{11}(t-1948)$

11d. $\quad y(2000) = 50 \cos \frac{2\pi}{11}(52)+60 = 52.9$

11e. $\quad 1948+k11 > 2000$

$\quad\quad\quad\quad k > 4.7, \ k = 5$

$\quad\quad 1948+5(11) = 2003$

1. $\text{per} = \dfrac{\pi}{\frac{1}{2}} = 2\pi$

3. $\text{per} = \dfrac{2\pi}{\pi} = 2$

5. $\text{per} = \dfrac{\pi}{\frac{1}{2}} = 2$

7. $\text{per} = \dfrac{\pi}{\frac{5}{6}} = \dfrac{6\pi}{5}$

9. $\text{per} = \dfrac{\pi}{2}$, asym: $\dfrac{\frac{\pi}{2}}{2} = \dfrac{\pi}{4} \pm n(\dfrac{\pi}{2})$

11. $\text{per} = \pi$, p.s. $= \dfrac{\pi}{4}$

    asym: $\pi + \dfrac{\pi}{4} = \dfrac{5\pi}{4}$, or $\dfrac{\pi}{4} + n\pi$

13. $\text{per} = \dfrac{\pi}{2}$, ps $= -\dfrac{\pi}{6}$

    asym: $\dfrac{\pi}{4} - \dfrac{\pi}{6} = \dfrac{\pi}{12}$, $\dfrac{\pi}{12} + n\dfrac{\pi}{2}$

15. $\text{per} = \dfrac{2\pi}{2} = \pi$, p.s. $= \dfrac{3\pi}{2}$

    asym: $\dfrac{3\pi}{2}$, $\dfrac{3\pi}{2} + n\pi$

17. $\text{per} = \pi$, p.s. $= \dfrac{\pi}{4}$

    asym: $\dfrac{\pi}{2} + \dfrac{\pi}{4} = \dfrac{3\pi}{4} + n\pi$

25. $\cot \theta = \dfrac{x}{20}$

    $x = 20 \cot \theta$

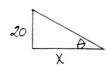

Section 8.6   Inverse Trigonometric Functions

1. $\text{Arcsin } \dfrac{1}{2} = y$

    $\sin y = \dfrac{1}{2}$

    $y = \dfrac{\pi}{6}$

3. $\text{Tan}^{-1} 1 = y$

    $\tan y = 1$, $y = \dfrac{\pi}{4}$

5. $\text{Arccos}\left(\dfrac{-\sqrt{3}}{2}\right) = y$

    $\cos y = \dfrac{-\sqrt{3}}{2}$, $y = \dfrac{5\pi}{6}$

7. $\text{Sin}^{-1}\left(\dfrac{-\sqrt{2}}{2}\right) = y$

    $\sin y = \dfrac{-\sqrt{2}}{2}$, $y = \dfrac{-\pi}{4}$

9. $\text{Arccos}(-1) = y$

    $\cos y = -1$, $y = \pi$

11. $\text{Sec}^{-1}(-2) = y$

    $\sec y = -2$,

    $\cos y = -\dfrac{1}{2}$, $y = \dfrac{2\pi}{3}$

13. $\cot^{-1} 0 = y$

    $\cot y = 0,\ y = \frac{\pi}{2}$

15. $\text{Arcsec } 1 = y$

    $\sec y = 1$

    $\cos y = 1,\ y = 0$

17. $\text{Arccos}\left(\frac{-\sqrt{2}}{2}\right) = y$

    $\cos y = \frac{-\sqrt{2}}{2}$

    $y = \frac{3\pi}{4}$

19. $\text{Arctan}\left(-\sqrt{3}\right) = y$

    $\tan y = -\sqrt{3},\ y = -\frac{\pi}{3}$

    $r = \sqrt{1+9} = \sqrt{10}$

    $\sin \theta = \frac{-3}{\sqrt{10}}$

21. $\sin\left(\text{Tan}^{-1}(-3)\right)$

    Let $\theta = \text{Tan}^{-1}(-3)$

    $\tan \theta = -3$

23. $\sin\left(\text{Sin}^{-1} 1\right)$

    Let $\theta = \text{Sin}^{-1} 1$

    $\sin \theta = 1$

25. $\sin\left(\text{Tan}^{-1} 2\right)$

    Let $\theta = \text{Tan}^{-1} 2$

    $\tan \theta = 2$

    $r = \sqrt{2^2+1} = \sqrt{5}$

    $\sin \theta = \frac{2}{\sqrt{5}}$

27. $\cos\left(\text{Arcsin } \frac{1}{4}\right)$

    Let $\theta = \text{Arcsin } \frac{1}{4}$

    $\sin \theta = \frac{1}{4}$

    $x = \sqrt{4^2-1} = \sqrt{15}$

    $\cos \theta = \frac{\sqrt{15}}{4}$

29. Arcsin (sin 170°)

Let Arcsin (sin 170°) $= \theta$

$\sin \theta = \sin 170°$

but $-\frac{\pi}{2} < \theta < \frac{\pi}{2}$

$\theta = 10° = \frac{\pi}{18}$

31. $\text{Cos}^{-1} (\cos 0) = \theta$

$\cos \theta = \cos 0$

$\theta = 0$

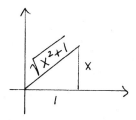

33. $\sin (\text{Arctan } x) = y$

Let Arctan $x = \theta$

$\tan \theta = x$

$r = \sqrt{x^2 + 1}$

$y = \sin \theta = \dfrac{x}{\sqrt{x^2 + 1}}$

35. $\sin (\text{Cos}^{-1} x^2) = y$

Let $\cos^{-1} x^2 = \theta$

$\cos \theta = x^2$

$y = \sin \theta = \sqrt{1 - x^2}$

37. $\cos \left(\text{Arcsin } (x-4) \right) = y$

Let $\theta = \text{Arcsin } (x-4)$

$(x-4) = \sin \theta$

$y = \cos \theta = \sqrt{1 - (x-4)^2}$

$= \sqrt{8x - x^2 - 15}$

39. Let $x = 1$, $y = 0$

Arccos 1 + Arccos 0 $= 0 + \frac{\pi}{2}$

Arccos $(1+0) = $ Arccos 1 $= 0$

41. $y = \text{Arcsin } x$

For inverse fcn, $x = \text{Arcsin } y$

$y = \sin x$

43. $y = 3 \, \mathrm{Cos}^{-1} \, x$

For inverse fcn, $x = 3 \, \mathrm{Cos}^{-1} \, y$

$\frac{x}{3} = \mathrm{Cos}^{-1} \, y, \; y = \cos \frac{x}{3}$

45. $\mathrm{Arcsin} \, (\sin x) = y$

$\sin x = \sin y \quad \text{and} \quad \frac{-\pi}{2} \le y \le \frac{\pi}{2}$

$x = y \quad \text{and} \quad \frac{-\pi}{2} \le y \le \frac{\pi}{2}$

47. $\sin \left(\frac{\pi}{2} + \mathrm{Arcsin} \, x \right) = \cos \, (\mathrm{Arcsin} \, x)$

Let $\theta = \mathrm{Arcsin} \, x$

$\sin \theta = x$

$\cos \theta = \sqrt{1 - x^2}$

49. $\sin \left(\mathrm{Arccos} \, \frac{u}{2} \right)$

Let $\theta = \mathrm{Arccos} \, \frac{u}{2}$

$\cos \theta = \frac{u}{2}$

$\sin \theta = \frac{\sqrt{4 - u^2}}{2}$

Chapter 8   Review

1a. $f(0) = 2 \sin \, (0 - 1) = -1.6829$

1b. $f\!\left(\frac{1}{3}\right) = 2 \sin \left(3(\frac{1}{3}) - 1 \right) = 0$

1c. $f(\frac{2\pi}{3}) = 2 \sin \left(3(\frac{2\pi}{3}) - 1 \right) = -1.6829$

3. $f(x) = 2 \sin \, (3x - 1) = 1 \quad \text{per} = \frac{2\pi}{3}$

$\sin \, (3x - 1) = \frac{1}{2}$

$3x - 1 = \frac{\pi}{6}, \frac{5\pi}{6}$

$x = \frac{\pi}{18} + \frac{1}{3} \pm \left(\frac{2\pi}{3} \right) n$

or

$= \frac{5\pi}{18} + \frac{1}{3} \pm \left(\frac{2\pi}{3} \right) n$

5.   per $= 2\pi$, amp $= .4$

7.   per $= \frac{2\pi}{7}$, amp $= 20$

9.   per $= \frac{2\pi}{2} = \pi$, amp $= .3$

    mean value $= 2$

11.   per $= \frac{2\pi}{2} = \pi$, amp $= 1$

    p.s. $= \dfrac{\frac{-\pi}{8}}{2} = -\frac{\pi}{16}$

13.   per $= 2\pi$, amp $= 1$

    p.s. $= \frac{\pi}{3}$, mean value $= 2$

15.   per $= \dfrac{\pi}{\frac{1}{2}} = 2\pi$, p.s. $= \dfrac{\frac{-\pi}{8}}{\frac{1}{2}} = -\frac{\pi}{4}$

    asym: $-\frac{\pi}{4} + n(2\pi)$

17.   per $= \dfrac{2\pi}{\frac{1}{3}} = 6\pi$

    p.s. $= \pi$, asym: $-\frac{\pi}{2} + n(3\pi)$

19.   amp $= 9$, per $= \frac{2\pi}{3}$

21. $x + 3 = \frac{\pi}{2}, \frac{3\pi}{2}, \frac{(2n+1)}{2}\pi$

    $x = \frac{\pi}{2} - 3, \frac{3\pi}{2} - 3, \left(\frac{2n+1}{2}\right)\pi - 3$

23. $\operatorname{Arcsin}\left(-\frac{1}{2}\right) = \theta$

    $\sin\theta = -\frac{1}{2}, \theta = -\frac{\pi}{6}$

25. $\sin\left(\operatorname{Cos}^{-1}\frac{1}{3}\right)$

    Let $\operatorname{Cos}^{-1}\frac{1}{3} = \theta$

        $\cos\theta = \frac{1}{3}$

        $\sin\theta = \frac{\sqrt{8}}{3}$

27. $\cos\left(\operatorname{Cos}^{-1}.3\right) = y$

    Let $\theta = \operatorname{Cos}^{-1}.3$

    $y = \cos\theta = .3$

29. $\sin\left(\mathrm{Tan}^{-1}\ 2\right)$

Let $\theta = \mathrm{Tan}^{-1}\ 2$

$\tan\theta = 2$

$r = \sqrt{2^2 + 1} = \sqrt{5}$

$\sin\theta = \dfrac{2}{\sqrt{5}}$

31. $\tan\left[\mathrm{Cot}^{-1}\left(-2\right)\right]$

Let $\theta = \mathrm{Cot}^{-1}\left(-2\right)$

$\cot\theta = -2$

$\tan\theta = -\dfrac{1}{2}$

Chapter 9 Trigonometric Identities and Equations

Section 9.1   Fundamental Trigonometric Relations

1. $\cos \theta + \tan \theta \sin \theta$

$$= \cos \theta + \frac{\sin \theta}{\cos \theta} \sin \theta$$

$$= \frac{\cos^2 \theta + \sin^2 \theta}{\cos \theta} = \frac{1}{\cos \theta} = \sec \theta$$

3. $(\tan x + \cot x) \sin x$

$$= \left(\frac{\sin x}{\cos x} + \frac{\cos x}{\sin x} \right) \sin x = \frac{(\sin^2 x + \cos^2 x)}{\cos x \, \sin x} \cdot \sin x = \frac{1}{\cos x} = \sec x$$

5. $\dfrac{\tan x \,(1 + \cot^2 x)}{1 + \tan^2 x} = \dfrac{\tan x \csc^2 x}{\sec^2 x} = \dfrac{\sin x}{\cos x} \cdot \dfrac{1}{\sin^2 x} \cdot \cos^2 x = \dfrac{\cos x}{\sin x} = \cot x$

7. $\cos x \cdot \csc x = \cos x \cdot \dfrac{1}{\sin x} = \cot x$

9. $(\cos^2 x - 1)(\tan^2 x + 1) = (-\sin^2 x)(\sec^2 x) = -\dfrac{\sin^2 x}{\cos^2 x} = -\tan^2 x$

11. $\dfrac{\sec x - \cos x}{\tan x} = \dfrac{\frac{1}{\cos x} - \cos x}{\frac{\sin x}{\cos x}} = \dfrac{1 - \cos^2 x}{\cos x} \cdot \dfrac{\cos x}{\sin x} = \dfrac{\sin^2 x}{\sin x} = \sin x$

13. $(\sin^2 x + \cos^2 x)^3 = 1^3 = 1$

15. $(\csc x - \cot x)^4 \,(\csc x + \cot x)^4 = (\csc^2 x - \cot^2 x)^4 = 1^4 = 1$

17. $\tan x \,(\sin x + \cot x \cos x) = \dfrac{\sin x}{\cos x}(\sin x + \dfrac{\cos x}{\sin x} \cdot \cos x) =$

$$= \dfrac{\sin^2 x + \cos^2 x}{\cos x} = \dfrac{1}{\cos x} = \sec x$$

19. $\dfrac{\tan x \sin x}{\sec^2 x - 1} = \dfrac{\tan x \sin x}{\tan^2 x} = \dfrac{\sin x}{\frac{\sin x}{\cos x}} = \cos x$

21. $\sqrt{a^2+x^2} = \sqrt{a^2 + a^2\tan^2\theta} = \sqrt{a^2(1+\tan^2\theta)} = a\sqrt{\sec^2\theta} = a\sec\theta$

$$\sec\theta = \frac{\sqrt{a^2+x^2}}{a}$$

$$\sin\theta = \frac{x}{\sqrt{a^2+x^2}}$$

23. $\dfrac{\sqrt{x^2-4}}{x} = \dfrac{\sqrt{4\sec^2\theta-4}}{2\sec\theta} = \dfrac{\sqrt{4(\sec^2\theta-1)}}{2\sec\theta} = \dfrac{2\sqrt{\tan^2\theta}}{2\sec\theta} = \dfrac{\tan\theta}{\sec\theta} =$

$\qquad = \dfrac{\sin\theta}{\cos\theta}\cdot\cos\theta = \sin\theta$

25. $\sqrt{3-5x^2} = \sqrt{3-5\cdot\frac{3}{5}\sin^2\theta} = \sqrt{3(1-\sin^2\theta)} = \sqrt{3}\sqrt{\cos^2\theta} = \sqrt{3}\cos\theta$

27. $\dfrac{\cos x - \cos^3 x}{\sin x - \sin^3 x} = \dfrac{\cos x\,(1-\cos^2 x)}{\sin x\,(1-\sin^2 x)} = \dfrac{\cos x\,\sin^2 x}{\sin x\,\cos^2 x} = \dfrac{\sin x}{\cos x} = \tan x$

29. $(\sec x - \tan x)^{-1} = \dfrac{1}{(\sec x - \tan x)}\dfrac{(\sec xc + \tan x)}{(\sec x + \tan x)} = \dfrac{\sec x + \tan x}{\sec^2 x - \tan^2 x} = \sec x + \tan x$

Section 9.2   Trigonometric Identities

1. $\sin x \cot x = \cos x$

$\cancel{\sin x}\dfrac{\cos x}{\cancel{\sin x}} = \cos x$

3. $\qquad \sec x \cot x = \csc x$

$\dfrac{1}{\cancel{\cos x}}\cdot\dfrac{\cancel{\cos x}}{\sin x} = \csc x$

$\dfrac{1}{\sin x} = \csc x$

5. $\sin^2 x\,(1+\cot^2 x) = 1$

$\sin^2 x \csc^2 x = 1$

$\cancel{\sin^2 x}\cdot\dfrac{1}{\cancel{\sin^2 x}} = 1$

7. $\csc x - \sin x = \cot x \cos x$

$\dfrac{1}{\sin x} - \sin x =$

$\dfrac{\cos^2 x}{\sin x} =$

$\dfrac{\cos x}{\sin x}\cdot\cos x = \cot x \cos x$

9. $\qquad (\sin^2 x - 1)(\cot^2 x - 1) = 1 - \csc^2 x$

$\sin^2 x \cot^2 x - \cot^2 x + \sin^2 x - 1 =$

$\cos^2 x - \cot^2 x + \sin^2 x - 1 =$

$1 - \cot^2 x - 1 =$

$1 - (\cot^2 x + 1) = 1 - \csc^2 x$

11.
$$\frac{2 + \sec x}{\csc x} - 2 \sin x = \tan x$$

$$\frac{2 + \sec x - 2 \sin x \csc x}{\csc x} =$$

$$\frac{2 + \sec x - 2}{\csc x} = \tan x$$

13.
$$\frac{\sin x}{1 - \cos x} = \csc x + \cot x$$

$$\frac{\sin x (1 + \cos x)}{(1 - \cos x)(1 + \cos x)} =$$

$$\frac{\sin x (1 + \cos x)}{1 - \cos^2 x} =$$

$$\frac{1 + \cos x}{\sin x} = \csc x + \cot x$$

15.
$$\frac{\cot x + 1}{\cot x - 1} = -\frac{\tan x + 1}{\tan x - 1}$$

$$\frac{\frac{\cos x}{\sin x} + 1}{\frac{\cos x}{\sin x} - 1} =$$

$$\frac{\cos x + \sin x}{\sin x} \cdot \frac{\sin x}{\cos x - \sin x} =$$

$$\frac{1 + \frac{\sin x}{\cos x}}{1 - \frac{\sin x}{\cos x}} = -\frac{\tan x + 1}{\tan x - 1}$$

17.
$$\frac{1 + \sec x}{\sin x + \tan x} = \csc x$$

$$\frac{1 + \frac{1}{\cos x}}{\sin x + \frac{\sin x}{\cos x}} =$$

$$\frac{\cos x + 1}{\cos x} \cdot \frac{\cos x}{\sin x \cos x + \sin x} =$$

$$\frac{(\cos x + 1)}{\sin x (\cos x + 1)} = \csc x$$

19.
$$\frac{1 - \sin x}{1 + \sin x} = (\sec x - \tan x)^2$$

$$\frac{(1 - \sin x)(1 - \sin x)}{(1 + \sin x)(1 - \sin x)} =$$

$$\frac{(1 - \sin x)^2}{1 - \sin^2 x} =$$

$$\frac{1 - 2 \sin x + \sin^2 x}{\cos^2 x} =$$

$$\sec^2 x - 2 \sec x \tan x + \tan^2 x = (\sec x - \tan x)^2$$

21. $(\sin^2 x + \cos^2 x)^4 = 1$

$$1^4 = 1$$

23.
$$\frac{\tan x + \cot x}{\tan x - \cot x} = \frac{\sec^2 x}{\tan^2 x - 1}$$

$$\frac{\dfrac{\sin x}{\cos x} + \dfrac{\cos x}{\sin x}}{\dfrac{\sin x}{\cos x} - \dfrac{\cos x}{\sin x}} =$$

$$\frac{\sin^2 x + \cos^2 x}{\cancel{\cos x \sin x}} \cdot \frac{\cancel{\cos x \sin x}}{\sin^2 x - \cos^2 x} =$$

$$\frac{1}{\sin^2 x - \cos^2 x} =$$

$$\frac{\dfrac{1}{\cos^2 x}}{\dfrac{\sin^2 x}{\cos^2 x} - 1} = \frac{\sec^2 x}{\tan^2 x - 1}$$

25.
$$\frac{\sin x}{\csc x (1 + \cot^2 x)} = \sin^4 x$$

$$\frac{\sin^2 x}{\csc^2 x} = \sin^4 x$$

27.
$$1 - \tan^4 x = 2 \sec^2 x - \sec^4 x$$

$$(1 + \tan^2 x)(1 - \tan^2 x) =$$

$$\sec^2 x\left(1 - (\sec^2 x - 1)\right) =$$

$$\sec^2 x(2 - \sec^2 x) = 2 \sec^2 x - \sec^4 x$$

29.
$$(\cot x + \csc x)^2 = \frac{1 + \cos x}{1 - \cos x}$$

$$\cot^2 x + 2 \cot x \csc x + \csc^2 x =$$

$$\frac{\cos^2 x}{\sin^2 x} + \frac{2 \cos x}{\sin^2 x} + \frac{1}{\sin^2 x} =$$

$$\frac{(\cos x + 1)^2}{\sin^2 x} =$$

$$\frac{(1 + \cos x)^2}{1 - \cos^2 x} =$$

$$\frac{(1 + \cos x)^2}{(1 - \cos x)(1 + \cos x)} = \frac{1 + \cos x}{1 - \cos x}$$

31. $\sec x \csc x - 2 \cos x \csc x = \tan x - \cot x$

$$\frac{1}{\cos x \sin x} - \frac{2 \cos x}{\sin x} =$$

$$\frac{1 - 2 \cos^2 x}{\cos x \sin x} =$$

$$\frac{1 - \cos^2 x - \cos^2 x}{\cos x \sin x} =$$

$$\frac{\sin^2 x - \cos^2 x}{\cos x \sin x} = \tan x - \cot x$$

33.
$$\tan^2 x - \cot^2 x = \sec^2 x - \csc^2 x$$

$$\sec^2 x - 1 - (\csc^2 x - 1) = \sec^2 x - \csc^2 x$$

35. $(1 - \sin^2 x)(1 + \tan^2 x) = 1$

$$(\cos^2 x)(\sec^2 x) = 1$$

37.
$$\frac{\cos x + \tan x}{\sin x \cos x} = \csc x + \sec^2 x$$

$$\frac{\cos x + \frac{\sin x}{\cos x}}{\sin x \cos x} =$$

$$\frac{\cos^2 x + \sin x}{\cos x} \cdot \frac{1}{\cos x \sin x} =$$

$$\frac{\cos^2 x + \sin x}{\cos^2 x \sin x} =$$

$$\frac{1}{\sin x} + \frac{1}{\cos^2 x} = \csc x + \sec^2 x$$

39.
$$(\sin x - \cos x)^2 = 1 - 2 \cot x \sin^2 x$$

$$\sin^2 x - 2 \sin x \cos x + \cos^2 x =$$

$$1 - 2 \sin x \cos x \left(\frac{\sin x}{\sin x}\right) =$$

$$1 - 2 \sin^2 x \frac{\cos x}{\sin x} = 1 - 2 \cot x \sin^2 x$$

41.
$$\cos^2 x - \sin x \tan x = \cos x \cot x \sin x - \tan x \sin x$$

$$= \cos x \frac{\cos x}{\sin x}\sin x - \tan x \sin x$$

$$= \cos^2 x - \tan x \sin x$$

43.
$$\tan x - \cot x = -\frac{\cos x - \sin^2 x \sec x}{\sin x}$$

$$= -\frac{\cos x}{\sin x} + \frac{\sin^2 x \sec x}{\sin x}$$

$$= -\cot x + \sin x \sec x$$

$$= -\cot x + \sin x \cdot \frac{1}{\cos x}$$

45.
$$\csc^4 x + \cot^4 x = 1 + 2 \csc^2 x \cot^2 x$$

$$= 1 + 2 \cdot \frac{1}{\sin^2 x} \cdot \frac{\cos^2 x}{\sin^2 x}$$

$$= \frac{\sin^4 x + 2 \cos^2 x}{\sin^4 x}$$

$$= \frac{(1 - \cos^2 x)^2 + 2 \cos^2 x}{\sin^4 x}$$

$$= \frac{1 - 2\cos^2 x + \cos^4 x + 2\cos^2 x}{\sin^4 x}$$

$$\csc^4 x + \cot^4 x = \frac{1}{\sin^4 x} + \frac{\cos^4 x}{\sin^4 x}$$

47. $(1 + \cos x)^2 = \sin^2 x \dfrac{(\sec x + 1)}{(\sec x - 1)}$

$$= \frac{\sin^2 x (\sec x + 1)^2}{\sec^2 x - 1}$$

$$= \frac{\sin^2 x (\sec x + 1)^2}{\tan^2 x}$$

$$= \frac{\cancel{\sin^2} x \cos^2 x}{\cancel{\sin^2} x}(\sec x + 1)^2$$

$$= \cos^2 x (\sec^2 x + 2 \sec x + 1)$$

$$(1 + \cos x)^2 = 1 + 2 \cos x + \cos^2 x$$

49. $(\sec x + \tan x)^2 = \dfrac{\sec x + \tan x}{\sec x - \tan x}$

$$= \frac{(\sec x + \tan x)^2}{\sec^2 x - \tan^2 x}$$

$$(\sec x + \tan x)^2 = \frac{(\sec x + \tan x)^2}{\cancel{\tan^2 x} + 1 - \cancel{\tan^2 x}}$$

51. $\dfrac{\csc x}{\csc x - \tan x} = \dfrac{\cos x}{\cos x - \sin^2 x}$

$$\frac{\dfrac{1}{\sin x}}{\dfrac{1}{\sin x} - \dfrac{\sin x}{\cos x}} =$$

$$\frac{1}{\cancel{\sin x}} \cdot \frac{\cancel{\sin x} \cos x}{\cos x - \sin^2 x} = \frac{\cos x}{\cos x - \sin^2 x}$$

53. $\dfrac{1}{1 - \sin x} - \dfrac{1}{1 + \sin x} = 2 \tan x \sec x$

$$\frac{1 + \sin x - (1 - \sin x)}{1 - \sin^2 x} =$$

$$\frac{2 \sin x}{\cos^2 x} =$$

$$2\frac{\sin x}{\cos x} \cdot \frac{1}{\cos x} = 2 \tan x \sec x$$

55.

$$\sec^4 x - \tan^4 x = \frac{1 + \sin^2 x}{\cos^2 x}$$

$$(\sec^2 x - \tan^2 x)(\sec^2 x + \tan^2 x) =$$

$$1 \cdot (\sec^2 x + \tan^2 x) =$$

$$\frac{1}{\cos^2 x} + \frac{\sin^2 x}{\cos^2 x} = \frac{1 + \sin^2 x}{\cos^2 x}$$

57.

$$(\csc x - \cot x)(\sec x + 1) = \tan x$$

$$\csc x \sec x + \csc x - \cot x \sec x - \cot x =$$

$$\frac{1}{\sin x \cos x} + \frac{1}{\sin x} - \frac{\cos x}{\sin x} \cdot \frac{1}{\cos x} - \frac{\cos x}{\sin x} =$$

$$\frac{1 - \cos^2 x}{\sin x \cos x} =$$

$$\frac{\sin^2 x}{\sin x \cos x} = \tan x$$

59.

$$(1 + \sin x)(\sec x - \tan x) = \cos x$$

$$\sec x - \tan x + \sin x \sec x - \sin x \tan x =$$

$$\frac{1}{\cos x} - \frac{\sin x}{\cos x} + \frac{\sin x}{\cos x} - \frac{\sin^2 x}{\cos x} =$$

$$\frac{1 - \sin^2 x}{\cos x} =$$

$$\frac{\cos^2 x}{\cos x} = \cos x$$

61.

$$\sec x \csc x - 2 \cos x \csc x = \tan x - \cot x$$

$$\frac{1}{\cos x \sin x} - \frac{2 \cos x}{\sin x} =$$

$$\frac{1 - 2 \cos^2 x}{\cos x \sin x} =$$

$$\frac{\sin^2 x + \cos^2 x - 2 \cos^2 x}{\cos x \sin x} =$$

$$\frac{\sin^2 x}{\cos x \sin x} - \frac{\cos^2 x}{\cos x \sin x} = \tan x - \cot x$$

63.
$$\frac{\sin x}{1 + \cos x} + \frac{1 + \cos x}{\sin x} = 2 \csc x$$

$$\frac{\sin x (\sin x) + (1 + \cos x)^2}{\sin x (1 + \cos x)} =$$

$$\frac{\sin^2 x + 1 + 2 \cos x + \cos^2 x}{\sin x (1 + \cos x)} =$$

$$\frac{2\cancel{(1 + \cos x)}}{\sin x \cancel{(1 + \cos x)}} = 2 \csc x$$

65.
$$\frac{\csc x}{\tan x + \cot x} = \cos x$$

$$\frac{\dfrac{1}{\sin x}}{\dfrac{\sin x}{\cos x} + \dfrac{\cos x}{\sin x}} =$$

$$\frac{1}{\cancel{\sin x}} \cdot \frac{\cancel{\sin x} \cos x}{\sin^2 x + \cos^2 x} = \cos x$$

67.
$$\frac{1 + \cos x}{1 - \cos x} = (\csc x + \cot x)^2$$

$$= \csc^2 x + 2 \csc x \cot x + \cot^2 x$$

$$= \frac{1 + 2 \cos x + \cos^2 x}{\sin^2 x}$$

$$= \frac{(1 + \cos x)^2}{\sin^2 x}$$

$$= \frac{(1 + \cos x)^2}{1 - \cos^2 x}$$

$$\frac{1 + \cos x}{1 - \cos x} = \frac{(1 + \cos x)^{\cancel{2}}}{\cancel{(1 + \cos x)}(1 - \cos x)}$$

69.
$$\frac{\cos^2 x}{1 - \sin x + \cos^2 x} = \frac{1 + \sin x}{2 + \sin x}$$

$$\frac{1 - \sin^2 x}{1 - \sin x + 1 - \sin^2 x} =$$

$$\frac{1 - \sin^2 x}{2 - \sin x - \sin^2 x} =$$

$$\frac{\cancel{(1 - \sin x)}(1 + \sin x)}{(2 + \sin x)\cancel{(1 - \sin x)}} = \frac{1 + \sin x}{2 + \sin x}$$

71.
$$(1 + \cot x)^2 = \csc^2 x(1 + 2 \cos x \sin x)$$

$$1 + 2 \cot x + \cot^2 x =$$

$$1 + 2\frac{\cos x}{\sin x} + \frac{\cos^2 x}{\sin^2 x} =$$

$$\frac{\sin^2 x + 2 \cos x \sin x + \cos^2 x}{\sin^2 x} =$$

$$\frac{1 + 2 \cos x \sin x}{\sin^2 x} = \csc^2 x(1 + 2 \cos x \sin x)$$

73. $(\cos x + \sin x + \tan x)^2 = \sec^2 x(1 + 2 \sin^2 x \cos x) + 2 \sin x(1 + \cos x)$

$$(\cos x + \sin x)^2 + 2(\cos x + \sin x) \tan x + \tan^2 x =$$

$$= \cos^2 x + 2 \cos x \sin x + \sin^2 x + 2 \cos x \tan x + 2 \sin x \tan x + \tan^2 x =$$

$$= 1 + 2\frac{\sin^2 x}{\cos x} + \frac{\sin^2 x}{\cos^2 x} + 2 \sin x(\cos x + 1) =$$

$$= \frac{\cos^2 x + 2 \sin^2 x \cos x + \sin^2 x}{\cos^2 x} + 2 \sin x(\cos x+1) =$$

$$= \frac{1}{\cos^2 x}(1 + 2 \sin^2 x \cos x) + 2 \sin x(\cos x + 1) =$$

$$= \sec^2 x(1 + 2 \sin^2 x \cos x) + 2 \sin x(1 + \cos x)$$

75. $(1 + \sin^2 x)^4 = 16 - 32 \cos^2 x + 24 \cos^4 x - 8 \cos^6 x + \cos^8 x$

$$1 + 4 \sin^2 x + 6 \sin^4 x + 4 \sin^6 x + \sin^8 x =$$

$$1 + 4 (1 - \cos^2 x) + 6 (1 - \cos^2 x)^2 + 4 (1 - \cos^2 x)^3 + (1 - \cos^2 x)^4 =$$

$$1 + 4 - 4 \cos^2 x + 6 - 12 \cos^2 x + 6 \cos^4 x + 4 - 12 \cos^2 x + 12 \cos^4 x - 4 \cos^6 x$$

$$+ 1 - 4 \cos^2 x + 6 \cos^4 x - 4 \cos^6 x + \cos^8 x =$$

$$= 16 - 32 \cos^2 x + 24 \cos^4 x - 8 \cos^6 x + \cos^8 x$$

77. $\dfrac{\tan x - \tan y}{\cot x - \cot y} = \dfrac{\dfrac{\sin x}{\cos x} - \dfrac{\sin y}{\cos y}}{\dfrac{\cos x}{\sin x} - \dfrac{\cos y}{\sin y}} =$

$$\dfrac{\dfrac{-(-\sin x \cos y + \sin y \cos x)}{\cos x \cos y}}{\dfrac{\cos x \sin y - \sin x \cos y}{\sin x \sin y}} = -\dfrac{\sin x \sin y}{\cos x \cos y}$$

$\dfrac{1 - \tan x \tan y}{1 - \cot x \cot y} = \dfrac{1 - \dfrac{\sin x}{\cos x}\dfrac{\sin y}{\cos y}}{1 - \dfrac{\cos x}{\sin x}\dfrac{\cos y}{\sin y}} =$

$$\dfrac{\dfrac{-(-\cos x \cos y + \sin x \sin y)}{\cos x \cos y}}{\dfrac{\sin x \sin y - \cos x \cos y}{\sin x \sin y}} = -\dfrac{\sin x \sin y}{\cos x \cos y}$$

79. $\tan x + \cot x = \sec x \csc x$

$\dfrac{\sin x}{\cos x} + \dfrac{\cos x}{\sin x} =$

$\dfrac{\sin^2 x + \cos^2 x}{\cos x \sin x} =$

$\dfrac{1}{\cos x \sin x} = \sec x \csc x$

81. $(2 \cos x - \sin x)^2 + (2 \sin x + \cos x)^2 = 5$

$4 \cos^2 x - \cancel{4 \cos x \sin x} + \sin^2 x + 4 \sin^2 x + \cancel{4 \sin x \cos x} + \cos^2 x =$

$$4 (\sin^2 x + \cos^2 x) + (\sin^2 x + \cos^2 x) = 5$$

83. $\cos t = \sqrt{\cos^2 t}$

Let $t = \dfrac{3\pi}{4}$

$\cos \dfrac{3\pi}{4} = \dfrac{-\sqrt{2}}{2}, \sqrt{\cos^2 \dfrac{3\pi}{4}} = \sqrt{\dfrac{2}{4}} = \dfrac{\sqrt{2}}{2}$

85. $1 = \sec (\cos x)$

Let $x = 0$

$\sec (\cos 0) = \sec 1 = 1.8508$

87. $\cos^2 x = \dfrac{1 - \sin x}{2}$

Let $x = 0$

$\cos^2 0 = 1, \dfrac{1 - \sin 0}{2} = \dfrac{1}{2}$

89. $\sin \frac{x}{2} = \frac{1}{2} \sin x$

Let $x = \frac{\pi}{2}$

$\sin \frac{\pi}{4} = \frac{\sqrt{2}}{2}, \frac{1}{2} \sin \frac{\pi}{2} = \frac{1}{2}$

91. $\sin (x + \pi) = \frac{1}{2} \sin x$

Let $x = \frac{\pi}{4}$

$\sin (\frac{\pi}{4} + \pi) = \sin \frac{5\pi}{4} = -\frac{\sqrt{2}}{2}$

$\sin \frac{\pi}{4} = \frac{\sqrt{2}}{2}$

93. $\cos x^2 = \cos^2 x$

Let $x = 2$

$\cos 2^2 = \cos 4 = -.6536$

$\cos^2 x = (\cos 2)^2 = .1732$

Section 9.3 Trigonometric Equations

7. $\sin 2x = -\frac{1}{2}$

$2x = \frac{7\pi}{6}, \frac{11\pi}{6}$

$x = \frac{7\pi}{12}, \frac{11\pi}{12}$

9. $|\cos x| = \frac{\sqrt{2}}{2}$

$\cos x = \frac{\sqrt{2}}{2}$ $\cos x = \frac{-\sqrt{2}}{2}$

$x = \frac{\pi}{4}, \frac{7\pi}{4}$ $x = \frac{3\pi}{4}, \frac{5\pi}{4}$

11. $2 \sin x + 1 = 0$

$\sin x = -\frac{1}{2}$

$x = \frac{7\pi}{6}, \frac{11\pi}{6}$

13. $\cos 3x = 1$

$3x = 0$

$x = 0$

15. $\cos^2 x + 2 \cos x + 1 = 0$

$(\cos x + 1)^2 = 0$

$\cos x = -1$

$x = \pi$

17. $2 \sin^2 x - \sin x = 0$

$\sin x (2 \sin x - 1) = 0$

$\sin x = 0$ $2 \sin x - 1 = 0$

$x = 0$ $\sin x = \frac{1}{2}$

$x = \frac{\pi}{6}, \frac{5\pi}{6}$

19. $\sec^2 x + 1 = 0$

    $\sec^2 x = -1$, No solution

21. $\cos x = \sin x$

    $x = \frac{\pi}{4}, \frac{5\pi}{4}$

23. $\sec^2 x - 2 = \tan^2 x$

    $\sec^2 x - 2 = \sec^2 x - 1$

        $-2 = -1$, No solution

25. $\qquad 2\cos^2 x - \sin x = 1$

    $2(1 - \sin^2 x) - \sin x = 1$

      $1 - 2\sin^2 x - \sin x = 0$

    $(2\sin x - 1)(\sin x + 1) = 0$

    $2\sin x - 1 = 0 \qquad \sin x + 1 = 0$

        $\sin x = \frac{1}{2} \qquad\quad \sin x = -1$

          $x = \frac{\pi}{6}, \frac{5\pi}{6} \qquad\quad x = \frac{3\pi}{2}$

27. $\sin^2 x - 2\sin x + 1 = 0$

    $(\sin x - 1)^2 = 0$

        $\sin x = 1$

           $x = \frac{\pi}{2}$

29. $\tan^2 x - \tan x = 0$

    $\tan x (\tan x - 1) = 0$

      $\tan x = 0 \qquad\quad \tan x = 1$

        $x = 0, \pi \qquad\quad x = \frac{\pi}{4}, \frac{5\pi}{4}$

31. $\qquad\quad \sin x \tan^2 x = \sin x$

    $\sin x (\tan^2 x - 1) = 0$

      $\sin x = 0 \qquad\quad \tan^2 x = 1$

        $x = 0, \pi \qquad\quad \tan x = \pm 1$

                  $x = \frac{\pi}{4}, \frac{3\pi}{4}, \frac{5\pi}{4}, \frac{7\pi}{4}$

33. $\tan x + \sec x = 1$

    $(\tan x + \sec x)^2 = 1$

      $\tan^2 x + 2\tan x \sec x + \sec^2 x = 1$

      $\tan^2 x + 2\tan x \sec x + \cancel{\sec^2 x} = \cancel{\sec^2 x} - \tan^2 x$

        $2\tan^2 x + 2\tan x \sec x = 0$

        $2\tan x (\tan x + \sec x) = 0$

(*problem #33, continued*)

$$\tan x = 0 \qquad\qquad \tan x + \sec x = 0$$
$$x = 0, \pi \qquad\qquad \frac{\sin x}{\cos x} + \frac{1}{\cos x} = 0$$
$$\frac{1}{\cos x}(\sin x - 1) = 0$$
$$\sin x = -1$$
$$x = \frac{3\pi}{2}$$

35.
$$\cos x + 1 = \sin x$$
$$\cos^2 x + 2\cos x + 1 = \sin^2 x$$
$$\cos^2 x + 2\cos x + 1 = 1 - \cos^2 x$$
$$2\cos^2 x + 2\cos x = 0$$
$$\cos x\,(\cos x + 1) = 0$$

$$\cos x = 0 \qquad\qquad \cos x = -1$$
$$x = \frac{\pi}{2}, \frac{3\pi}{2}\!\!\!\!\diagup \qquad\qquad x = \pi$$

$$\cos \frac{\pi}{2} + 1 \overset{?}{=} \sin \frac{\pi}{2} \qquad \text{yes}$$
$$\cos \frac{3\pi}{2} + 1 \overset{?}{=} \sin \frac{3\pi}{2},\ 1 = -1, \qquad \text{No}$$
$$\cos \pi + 1 \overset{?}{=} \sin \pi,\ 0 = 0, \qquad \text{yes}$$

37. $\csc^5 x - 4\csc x = 0$
$$\csc x\,(\csc^4 x - 4) = 0$$

$$\csc x = 0 \qquad\qquad \csc^4 x = 4$$
$$\text{No solution} \qquad\qquad \csc x = \pm\sqrt{2}$$
$$\sin x = \frac{\pm 1}{\sqrt{2}}$$
$$x = \frac{\pi}{4}, \frac{3\pi}{4}, \frac{5\pi}{4}, \frac{7\pi}{4}$$

39. $2\cos^2 \theta = 1$
$$\cos^2 \theta = \frac{1}{2}$$
$$\cos \theta = \pm\sqrt{\frac{1}{2}}$$
$$\theta = 45°, 135°, 225°, 315°$$

41. $\csc^2 \theta = 1$
$$\csc \theta = \pm 1$$
$$\sin \theta = \pm 1$$
$$\theta = 90°, 270°$$

43. $(\sin \theta - 1)(2 \cos \theta + \sqrt{3}) = 0$

$\sin \theta = 1 \qquad \cos \theta = -\dfrac{\sqrt{3}}{2}$

$\theta = 90° \qquad \theta = 150°, 210°$

45. $2 \sin \theta \tan \theta + \sqrt{3} \tan \theta = 0$

$\tan \theta \, (2 \sin \theta + \sqrt{3}) = 0$

$\theta = 0°, 180° \qquad \theta = 240°, 300°$

47. $\tan^2 \theta - \tan \theta = 0$

$\tan \theta \, (\tan \theta - 1) = 0$

$\tan \theta = 0 \qquad \tan \theta = 1$

$\theta = 0°, 180° \qquad \theta = 45°, 225°$

49. $2 \cos^2 \theta + 7 \sin \theta = 5$

$2 \, (1 - \sin^2 \theta) + 7 \sin \theta - 5 = 0$

$-2 \sin^2 \theta + 7 \sin \theta - 3 = 0$

$(2 \sin \theta - 1)(\sin \theta - 3) = 0$

$\sin \theta = \dfrac{1}{2} \qquad\qquad \sin \theta = 3$

$\theta = 30°, 150° \qquad \text{No solution}$

51. $\sin \theta + \sqrt{3} \cos \theta = 0$

$\sin \theta = -\sqrt{3} \cos \theta$

$\tan \theta = -\sqrt{3}$

$\theta = 120°, 300°$

53. $\sin^2 x + 2 \cos^2 x = 1.7$

$1 - \cos^2 x + 2 \cos^2 x = 1.7$

$1 + \cos^2 x = 1.7$

$\cos^2 x = .7$

$\cos x = \pm .8366$

$x = .5796, 2.562, 3.721, 5.704$

55. $3 \sin^2 x + \sin x = 0$

$\sin x \, (3 \sin x + 1) = 0$

$\sin x = 0 \qquad \sin x = \dfrac{-1}{3}$

$x = 0, \pi \qquad x = 5.943, 3.481$

57. $2 \sin x - \cot x \cos x = 1$

$2 \sin x - \dfrac{\cos^2 x}{\sin x} = 1$

$2 \sin^2 x - \cos^2 x = \sin x$

$2 \sin^2 x - (1 - \sin^2 x) = \sin x$

$3 \sin^2 x - \sin x - 1 = 0$

$\sin x = \dfrac{1 \pm \sqrt{1 + 12}}{6} = .7675, -.4343$

$\sin x = .7675 \qquad \sin x = -.4343$

$x = .875, 2.267 \qquad x = 5.83, 3.591$

59. $\sin x - \csc x + 1 = 0$

$$\sin x - \frac{1}{\sin x} + 1 = 0$$

$$\sin^2 x - 1 + \sin x = 0$$

$$\sin x = \frac{-1 \pm \sqrt{1+4}}{2} = -1.618, .618$$

$\sin x = -1.618 \qquad \sin x = .618$

No solution $\qquad x = .6662, 2.475$

61. $\sqrt{3} \cos \pi x + \sin \pi x = 0$

$$\sin \pi x = -\sqrt{3} \cos \pi x$$

$$\tan \pi x = -\sqrt{3}$$

$$\pi x = \frac{2\pi}{3}, \frac{5\pi}{3}$$

$$x = \frac{2}{3}, \frac{5}{3}$$

63. $V \sin (w_c t + \phi) = 0$

$$w_c t + \phi = 0, \pi, \ldots n\pi$$

$$t = \frac{n\pi - \phi}{w_c}$$

Section 9.4   Addition Formulas

1. $\cos \frac{\pi}{12} = \cos \left(\frac{\pi}{3} - \frac{\pi}{4} \right) = \cos \frac{\pi}{3} \cos \frac{\pi}{4} + \sin \frac{\pi}{3} \sin \frac{\pi}{4} =$

$$= \frac{1}{2}\left(\frac{\sqrt{2}}{2}\right) + \frac{\sqrt{3}}{2}\left(\frac{\sqrt{2}}{2}\right) = \frac{\sqrt{2} + \sqrt{6}}{4}$$

3. $\cos 195° = \cos (135° + 60°) = \cos 135° \cos 60° - \sin 135° \sin 60°$

$$= \frac{-\sqrt{2}}{2}\left(\frac{1}{2}\right) - \left(\frac{\sqrt{2}}{2}\right)\left(\frac{\sqrt{3}}{2}\right) = \frac{-(\sqrt{2} + \sqrt{6})}{4}$$

5. $\sin \frac{5\pi}{12} = \sin \left(\frac{8\pi}{12} - \frac{3\pi}{12} \right) = \sin \frac{2\pi}{3} \cos \frac{\pi}{4} - \sin \frac{\pi}{4} \cos \frac{2\pi}{3} =$

$$= \frac{\sqrt{3}}{2}\left(\frac{\sqrt{2}}{2}\right) - \frac{\sqrt{2}}{2}\left(\frac{-1}{2}\right) = \frac{\sqrt{6} + \sqrt{2}}{4}$$

7. $\sin \frac{7\pi}{12} = \sin \left(\frac{3\pi}{12} + \frac{4\pi}{12} \right) = \sin \frac{\pi}{4} \cos \frac{\pi}{3} + \cos \frac{\pi}{4} \sin \frac{\pi}{3} =$

$$= \frac{\sqrt{2}}{2}\left(\frac{1}{2}\right) + \frac{\sqrt{2}}{2}\left(\frac{\sqrt{3}}{2}\right) = \frac{\sqrt{2} + \sqrt{6}}{4}$$

9. $\cos (A + B) = \cos A \cos B - \sin A \sin B$

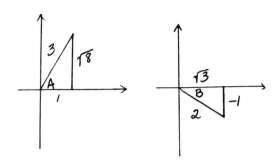

$$\cos A = \tfrac{1}{3}, \ \sin A = \frac{\sqrt{8}}{3}$$

$$\sin B = -\tfrac{1}{2}, \ \cos B = \frac{\sqrt{3}}{2}$$

$$\cos (A + B) = \tfrac{1}{3}\left(\frac{\sqrt{3}}{2}\right) - \left(\frac{\sqrt{8}}{3}\right)\left(-\tfrac{1}{2}\right) = \frac{\sqrt{6} + \sqrt{8}}{6}$$

11.

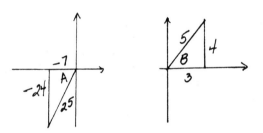

$$\tan A = \frac{24}{7}, \ \cos A = \frac{-7}{25}, \ \sin A = \frac{-24}{25}$$

$$\sec B = \tfrac{5}{3}, \ \cos B = \tfrac{3}{5}, \ \sin B = \tfrac{4}{5}$$

$$\cos (A + B) = \cos A \cos B - \sin A \sin B$$

$$= \left(\frac{-7}{25}\right)\left(\tfrac{3}{5}\right) - \left(-\frac{24}{25}\right)\left(\tfrac{4}{5}\right) = \frac{75}{5(25)} = \tfrac{3}{5}$$

13.

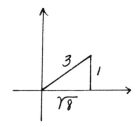

$\cos (A + B) = \cos A \cos B - \sin A \sin B$

$\dfrac{5}{6} = \cos A \left(\dfrac{\sqrt{8}}{3}\right) - \sin A \left(\dfrac{1}{3}\right)$

$\dfrac{5}{6} = \dfrac{\sqrt{8}}{3} \cos A - \dfrac{1}{3} \sqrt{1 - \cos^2 A}$

$5 - 2\sqrt{8} \cos A = -2\sqrt{1 - \cos^2 A}$

$25 - 20\sqrt{8} \cos A + 4\cdot 8 \cos^2 A = 4(1 - \cos^2 A)$

$36 \cos^2 A - 20\sqrt{8} \cos A + 21 = 0$

$\cos A = \dfrac{20\sqrt{8} \pm \sqrt{20^2(8) - 4(36)(21)}}{72} = \dfrac{5\sqrt{8} + \sqrt{11}}{18}, \dfrac{5\sqrt{8} - \sqrt{11}}{18}$ \quad (does not check)

15. $\tan A = \dfrac{-7}{24}$, $\sin A = \dfrac{7}{25}$, $\cos A = \dfrac{-24}{25}$

$\tan B = \dfrac{5}{12}$, $\sin B = \dfrac{-5}{13}$, $\cos B = \dfrac{-12}{13}$

$\sin (A + B) = \sin A \cos B + \cos A \sin B$

$= \dfrac{7}{25}\left(\dfrac{-12}{13}\right) + \left(\dfrac{-24}{25}\right)\left(\dfrac{-5}{13}\right) = \dfrac{36}{325}$

$\tan (A + B) = \dfrac{\tan A + \tan B}{1 - \tan A \tan B} = \dfrac{\dfrac{-7}{24} + \dfrac{5}{12}}{1 - \dfrac{7}{24}\left(\dfrac{5}{12}\right)} = \dfrac{36}{323}$

17. $\cos A = \frac{1}{3}$, $\sin A = \frac{-\sqrt{8}}{3}$, $\tan A = -\sqrt{8}$

  $\cos B = -\frac{1}{3}$, $\sin B = -\frac{\sqrt{8}}{3}$, $\tan B = \sqrt{8}$

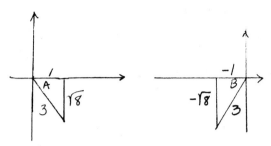

  $\sin (A + B) = \sin A \cos B + \cos A \sin B$

  $= -\frac{\sqrt{8}}{3}\left(-\frac{1}{3}\right) + \left(\frac{1}{3}\right)\left(\frac{-\sqrt{8}}{3}\right) = 0$

  $\tan (A + B) = \frac{\tan A + \tan B}{1 - \tan A \tan B} = 0$

19. $\frac{\tan 3x - \tan 2x}{1 + \tan 3x \tan 2x} = \tan (3x-2x) = \tan x$

21. $\frac{\tan (x + y) + \tan z}{1 - \tan (x+y) \tan z} = \tan (x+y+z)$

23. $\cos 7x \cos x - \sin 7x \sin x = \cos (7x + x) = \cos 8x$

25. $\cos \left(\frac{\pi}{3} - x\right) = \cos \frac{\pi}{3} \cos x + \sin \frac{\pi}{3} \sin x = \frac{1}{2} \cos x + \frac{\sqrt{3}}{2} \sin x$

27. $\cos \left(\frac{3\pi}{2} + x\right) = \cos \frac{3\pi}{2} \cos x - \sin \frac{3\pi}{2} \sin x = -(-1) \sin x = \sin x$

29. $\cos (x+y) + \cos (x-y) = \cos x \cos y - \sin x \sin y + \cos x \cos y + \sin x \sin y =$

  $= 2 \cos x \cos y$

31. $\tan \left(A+\frac{\pi}{2}\right) = \frac{\sin (A+\frac{\pi}{2})}{\cos (A+\frac{\pi}{2})} = \frac{\sin A \cos \frac{\pi}{2} + \cos A \sin \frac{\pi}{2}}{\cos A \cos \frac{\pi}{2} - \sin A \sin \frac{\pi}{2}} = \frac{\cos A}{-\sin A} = -\cot A$

33. $\cot (A+B) = \frac{\cos (A+B)}{\sin (A+B)} = \left(\frac{\cos A \cos B - \sin A \sin B}{\sin A \cos B + \cos A \sin B}\right) \cdot \left(\frac{\frac{1}{\sin A \sin B}}{\frac{1}{\sin A \sin B}}\right)$

  $= \frac{\cot A \cot B - 1}{\cot A + \cot B}$

35. $\sin (A+B) \sin (A-B)$

$= (\sin A \cos B + \sin B \cos A)(\sin A \cos B - \sin B \cos A)$

$= \sin^2 A \cos^2 B - \cos^2 A \sin^2 B$

$= \sin^2 A (1 - \sin^2 B) - (1 - \sin^2 A) \sin^2 B$

$= \sin^2 A - \cancel{\sin^2 A \sin^2 B} - \sin^2 B + \cancel{\sin^2 A \sin^2 B}$

$= \sin^2 A - \sin^2 B$

37. $\dfrac{\sin (A+B)}{\cos A \cos B} = \dfrac{\sin A \cos B + \sin B \cos A}{\cos A \cos B}$

$= \dfrac{\sin A \cancel{\cos B}}{\cos A \cancel{\cos B}} + \dfrac{\sin B \cancel{\cos A}}{\cancel{\cos A} \cos B} = \tan A + \tan B$

39. $\tan \alpha = \tan (\theta - \frac{\pi}{4}) = \dfrac{\tan \theta - \tan \frac{\pi}{4}}{1 + \tan \theta \tan \frac{\pi}{4}} = \dfrac{2-1}{1+2(1)} = \frac{1}{3}$

41. $I = I_0 \sin (\alpha t - \delta)$

$= I_0 (\sin \alpha t \cos \delta - \sin \delta \cos \alpha t)$

Section 9.5   Double-Angle and Half-Angle Formulas

3. $\sin^2 4x - \cos^2 4x = -(\cos^2 4x - \sin^2 4x) = -\cos 8x$

5. $\dfrac{2 \tan \frac{x}{6}}{1 - \tan^2 \frac{x}{6}} = \tan \frac{2x}{6} = \tan \frac{x}{3}$

7. $f(x) = \sin 2x \cos 2x = \frac{1}{2} \sin 4x$

per $= \frac{2\pi}{4} = \frac{\pi}{2}$, amp $= \frac{1}{2}$

max $= \frac{1}{2}$ at $\frac{\pi}{8}$

9. $f(x) = \sec x \csc x = \dfrac{1}{\cos x \sin x} = \dfrac{1}{\frac{1}{2} \sin 2x} = 2 \csc 2x$

per $= \frac{2\pi}{2}$, asym. at $2x = 0, \pi, 2\pi, x = 0, \frac{\pi}{2}, \pi$

11. $f(x) = \tan x + \cot x = \dfrac{\sin x}{\cos x} + \dfrac{\cos x}{\sin x} = \dfrac{\sin^2 x + \cos^2 x}{\sin x \cos x} = \dfrac{1}{\sin x \cos x} = 2 \csc 2x$

(See 9.5—9)

13.

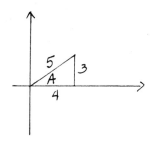

$\sin A = \frac{3}{5}$, $\cos A = \frac{4}{5}$, $\tan A = \frac{3}{4}$

$\sin 2A = 2 \sin A \cos A = 2\left(\frac{3}{5}\right)\left(\frac{4}{5}\right) = \frac{24}{25}$

$\cos 2A = \cos^2 A - \sin^2 A = \left(\frac{4}{5}\right)^2 - \left(\frac{3}{5}\right)^2 = \frac{7}{25}$

$\tan 2A = \dfrac{2 \tan A}{1 - \tan^2 A} = \dfrac{2\left(\frac{3}{4}\right)}{1 - \left(\frac{3}{4}\right)^2} = \dfrac{24}{7}$

15.

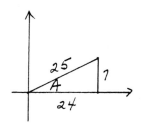

$\tan A = \frac{7}{24}$, $\sin A = \frac{-7}{25}$, $\cos A = \frac{-24}{25}$

$\sin 2A = 2 \sin A \cos A = 2\left(\frac{-7}{25}\right)\left(\frac{-24}{25}\right) = \frac{336}{625}$

$\cos 2A = \cos^2 A - \sin^2 A = \left(\frac{-24}{25}\right)^2 - \left(\frac{-7}{25}\right)^2 = \frac{527}{625}$

$\tan 2A = \dfrac{\sin 2A}{\cos 2A} = \dfrac{336}{527}$

17. $\sin \frac{\pi}{8} = \sin \frac{1}{2}\left(\frac{\pi}{4}\right) = +\sqrt{\dfrac{1 - \cos \frac{\pi}{4}}{2}} = +\sqrt{\dfrac{1 - \frac{\sqrt{2}}{2}}{2}} = \frac{1}{2}\sqrt{2 - \sqrt{2}}$

19. $\tan 157.5° = \tan \frac{1}{2}(315°) = \dfrac{\sin 315°}{1 + \cos 315°} = \dfrac{-\frac{\sqrt{2}}{2}}{1 + \frac{\sqrt{2}}{2}} = \dfrac{-\sqrt{2}}{2 + \sqrt{2}}$

21. $\cos \frac{\pi}{12} = \cos \frac{1}{2}\left(\frac{\pi}{6}\right) = +\sqrt{\dfrac{1 + \cos \frac{\pi}{6}}{2}} = \sqrt{\dfrac{1 + \frac{\sqrt{3}}{2}}{2}} = \frac{1}{2}\sqrt{2 + \sqrt{3}}$

23. $\cos 3x = 4 \cos^3 x - 3 \cos x$

$\cos (2x + x) =$

$\cos 2x \cos x - \sin 2x \sin x =$

$(\cos^2 x - \sin^2 x) \cos x - (2 \sin x \cos x) \sin x =$

$\cos^3 x - \cos x \sin^2 x - 2 \sin^2 x \cos x =$

$\cos^3 x - 3 (1 - \cos^2 x) \cos x =$

$\cos^3 x - 3 \cos x + 3 \cos^3 x = 4 \cos^3 x - 3 \cos x$

25. $\tan x + \cot x = 2 \csc 2x$

$\qquad = \dfrac{2}{\sin 2x}$

$\qquad = \dfrac{2}{2 \sin x \cos x}$

$\qquad = \dfrac{\sin^2 x + \cos^2 x}{\sin x \cos x}$

$\tan x + \cot x = \dfrac{\sin x}{\cos x} + \dfrac{\cos x}{\sin x}$

27. $\cot^2 \frac{x}{2} = \dfrac{\sec x + 1}{\sec x - 1}$

$\dfrac{\cos^2 \frac{x}{2}}{\sin^2 \frac{x}{2}} = \dfrac{\frac{1 + \cos x}{2}}{\frac{1 - \cos x}{2}} = \left(\dfrac{1 + \cos x}{1 - \cos x}\right) \cdot \dfrac{\sec x}{\sec x} = \dfrac{\sec x + 1}{\sec x - 1}$

29. $\sec^2 x = \dfrac{4 \sin^2 x}{\sin^2 2x}$

$\qquad = \dfrac{4 \sin^2 x}{(2 \sin x \cos x)^2}$

$\qquad = \dfrac{4 \sin^2 x}{4 \sin^2 x \cos^2 x}$

$\qquad = \sec^2 x$

31. $\qquad\qquad 2 \cot 2x \cos x = \csc x - 2 \sin x$

$\not{2}\left(\dfrac{1 - \tan^2 x}{\not{2} \tan x}\right)\cos x =$

$\left(\cos x - \dfrac{\sin^2 x}{\cos x}\right) \cdot \dfrac{\cos x}{\sin x} =$

$\dfrac{\cos^2 x - \sin^2 x}{\cancel{\cos x}} \cdot \dfrac{\cancel{\cos x}}{\sin x} =$

$\dfrac{1 - \sin^2 x - \sin^2 x}{\sin x} = \csc x - 2 \sin x$

33. $\tan^2 x + \cos 2x = 1 - \cos 2x \tan^2 x$

$\qquad\qquad = 1 - (2 \cos^2 x - 1) \tan^2 x$

$\qquad\qquad = 1 - 2 \cos^2 x \tan^2 x + \tan^2 x$

$\qquad\qquad = 1 - 2 \sin^2 x + \tan^2 x$

$\qquad\qquad = \cos 2x + \tan^2 x$

35. $\dfrac{\cos 3x}{\sec x} - \dfrac{\sin x}{\csc 3x} = \csc^2 2x - \sin^2 2x$

$\cos 3x \cos x - \sin x \sin 3x =$

$\cos (3x + x) =$

$\cos 4x = \cos^2 2x - \sin^2 2x$

37. $1 + \cot x \cot 3x = \dfrac{2 \cos 2x}{\cos 2x - \cos 4x}$

$\qquad\qquad = \dfrac{2 \cos (3x - x)}{\cos (3x - x) - \cos (3x + x)}$

$\qquad\qquad = \dfrac{2 \cos 3x \cos x + 2 \sin 3x \sin x}{\cancel{\cos 3x \cos x} + \sin 3x \sin x - \cancel{\cos 3x \cos x} + \sin 3x \sin x}$

$\cot x \cot 3x + 1 = \dfrac{2 (\cos 3x \cos x + \sin 3x \sin x)}{2 \sin 3x \sin x}$

39.
$$\sin 6x \tan 3x = 2 \sin^2 3x$$

$$2 \sin 3x \, \cancel{\cos 3x} \cdot \frac{\sin 3x}{\cancel{\cos 3x}} = 2 \sin^2 3x$$

41.
$$\cos 4x \sec^2 2x = 1 - \tan^2 2x$$

$$(\cos^2 2x - \sin^2 2x) \sec^2 2x =$$

$$1 - \frac{\sin^2 2x}{\cos^2 2x} = 1 - \tan^2 2x$$

43. $\sin 2\theta = \frac{5}{13}$, $\cos 2\theta = \frac{12}{13}$

$$\cos 2\theta = 2 \cos^2 \theta - 1$$

$$\frac{12}{13} = 2 \cos^2 \theta - 1$$

$$\cos^2 \theta = \frac{25}{26}$$

$$\cos \theta = \frac{5}{\sqrt{26}}, \ \sin \theta = \sqrt{1 - \frac{25}{26}} = \frac{1}{\sqrt{26}}$$

$$\tan \theta = \frac{1}{5}$$

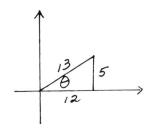

45. $\cos 2\theta = \frac{24}{25}$

$$\cos 2\theta = 2 \cos^2 \theta - 1$$

$$\cos^2 \theta = \frac{49}{50}$$

$$\cos \theta = \frac{7}{5\sqrt{2}}$$

47.
$$\sin 2x = \cos x$$

$$2 \sin x \cos x - \cos x = 0$$

$$\cos x (2 \sin x - 1) = 0$$

$$\cos x = 0 \qquad \sin x = \frac{1}{2}$$

$$x = \frac{\pi}{2}, \frac{3\pi}{2} \qquad x = \frac{\pi}{6}, \frac{5\pi}{6}$$

49.
$$\tan 2x = \tan x$$

$$\frac{2 \tan x}{1 - \tan^2 x} = \tan x$$

$$2 \tan x = \tan x - \tan^3 x$$

$$\tan^3 x + \tan x = 0$$

$$\tan x (\tan^2 x + 1) = 0$$

$$\tan x = 0 \qquad \tan^2 x = -1$$

$$x = 0, \pi, 2\pi \qquad \text{No solution}$$

51. $\sin 2x + \cos 2x = 0$

$$\sin 2x = -\cos 2x$$

$$\tan 2x = -1$$

$$2x = \frac{3\pi}{4}, \frac{7\pi}{4}, \frac{11\pi}{4}, \frac{15\pi}{4}$$

$$x = \frac{3\pi}{8}, \frac{7\pi}{8}, \frac{11\pi}{8}, \frac{15\pi}{8}$$

53. $2\left(\sin^2 2x - \cos^2 2x\right) = 1$

$$2\left(-\cos 4x\right) = 1$$

$$\cos 4x = -\frac{1}{2}$$

$$4x = \frac{2\pi}{3}, \frac{4\pi}{3}, \frac{8\pi}{3}, \frac{10\pi}{3}, \frac{14\pi}{3}, \frac{16\pi}{3}, \frac{20\pi}{3}, \frac{22\pi}{3}$$

$$x = \frac{\pi}{6}, \frac{\pi}{3}, \frac{2\pi}{3}, \frac{5\pi}{6}, \frac{7\pi}{6}, \frac{4\pi}{3}, \frac{5\pi}{3}, \frac{11\pi}{6}$$

55. $(\sin x + \cos x)^2 - \sin^2 2x = \sin^2 x + \cancel{2\sin x \cos x} + \cos^2 x - \cancel{2\sin x \cos x}$

$$= \sin^2 x + \cos^2 x = 1$$

57. $\sec^4 x - \tan^4 x - 2\tan^2 x$

$$= (\tan^2 x + 1)^2 - \tan^4 x - 2\tan^2 x$$

$$= \tan^4 x + 2\tan^2 x + 1 - \tan^4 x - 2\tan^2 x$$

$$= 1$$

59. $\cos 2x + \sin 2x \tan x$

$$= \cos^2 x - \sin^2 x + 2\sin x \cancel{\cos x}\left(\frac{\sin x}{\cancel{\cos x}}\right)$$

$$= \cos^2 x + \sin^2 x$$

$$= 1$$

65. $\tan^2 \theta = 2$

$$2 = \frac{2\tan \theta}{1 - \tan^2 \theta}$$

$$1 - \tan^2 \theta = \tan \theta$$

$$\tan^2 \theta + \tan \theta - 1 = 0$$

$$\tan \theta = \frac{-1 \pm \sqrt{1+4}}{2} = \frac{-1 + \sqrt{5}}{2}, \cancel{\frac{-1 - \sqrt{5}}{2}}$$

$$\tan^2 \theta + 1 = \sec^2 \theta$$

$$\sec^2 \theta = \left(\frac{-1 + \sqrt{5}}{2}\right)^2 + 1$$

$$= \frac{10 - 2\sqrt{5}}{4} = \frac{5 - \sqrt{5}}{2} = \frac{\sqrt{5}(\sqrt{5}-1)}{2}$$

$$\cos \theta = \sqrt{\frac{2}{\sqrt{5}(\sqrt{5}-1)}} \cdot \sqrt{\frac{\sqrt{5}+1}{\sqrt{5}+1}}$$

$$= \sqrt{\frac{2(\sqrt{5}+1)}{\sqrt{5}(5-1)}} = \sqrt{\frac{1+\sqrt{5}}{2\sqrt{5}}}$$

$$\sin \theta = \cos \theta \tan \theta$$

$$= \sqrt{\frac{1+\sqrt{5}}{2\sqrt{5}}} \cdot \frac{\sqrt{5}-1}{2} = \sqrt{\frac{\sqrt{5}-1}{2\sqrt{5}}}$$

67. $R = 32v_0{}^2 \sin \theta \cos \theta$

$R = 16v_0{}^2 \sin 2\theta$

$\sin 2\theta = \dfrac{R}{16v_0{}^2}$

Section 9.6   Sum and Product Formulas

1. $\sin 3\theta + \sin \theta = 2 \sin \frac{4\theta}{2} \cos \frac{2\theta}{2} = 2 \sin 2\theta \cos \theta$

3. $\sin 8x + \sin 2x = 2 \sin \frac{10x}{2} \cos \frac{6x}{2} = 2 \sin 5x \cos 3x$

5. $\cos 50° - \cos 30° = -2 \sin \frac{80°}{2} \sin \frac{20°}{2} = -2 \sin 40° \sin 10°$

7. $\sin \frac{3}{4} - \sin \frac{1}{4} = 2 \cos \frac{\left(\frac{3}{4}+\frac{1}{4}\right)}{2} \sin \frac{\left(\frac{3}{4}-\frac{1}{4}\right)}{2} = 2 \cos \frac{1}{2} \sin \frac{1}{4}$

9. $\cos x \sin \frac{x}{2} = \frac{1}{2}\left(\sin \left(x+\frac{x}{2}\right) - \sin \left(x-\frac{x}{2}\right)\right) = \frac{1}{2} \left(\sin \frac{3x}{2} - \sin \frac{x}{2}\right)$

11. $\cos 6x \cos 2x = \frac{1}{2} \left(\cos (6x+2x) + \cos (6x-2x)\right) = \frac{1}{2} (\cos 8x + \cos 4x)$

13. $\dfrac{\sin x + \sin y}{\cos x + \cos y} = \dfrac{2 \sin \left(\frac{x+y}{2}\right) \cos \left(\frac{x-y}{2}\right)}{2 \cos \left(\frac{x+y}{2}\right) \cos \left(\frac{x-y}{2}\right)} = \tan \left(\dfrac{x+y}{2}\right)$

15. $\dfrac{\cos 3x + \cos x}{\sin 3x + \sin y} = \dfrac{2 \cos 2x \cos x}{2 \sin 2x \cos x} = \cot 2x$

17. $\dfrac{\sin 2x + \sin 2y}{\cos 2x + \cos 2y} = \dfrac{2 \sin (x+y) \cos (x-y)}{2 \cos (x+y) \cos (x-y)} = \tan (x+y)$

19. $\dfrac{f(a+h) - f(a)}{h} = \dfrac{\cos (a+h) - \cos a}{h} = \dfrac{-2 \sin \left(\frac{2a+h}{2}\right) \sin \left(\frac{h}{2}\right)}{h}$

21. $\qquad \sin x - \sin 5x = 0$

$2 \cos \dfrac{6x}{2} \sin \left(\dfrac{-4x}{2}\right) = 0$

$\cos 3x = 0$	$\sin 2x = 0$
$3x = \dfrac{\pi}{2}, \dfrac{3\pi}{2}, \dfrac{5\pi}{2}$	$2x = 0, \pi, 2\pi$
$x = \dfrac{\pi}{6}, \dfrac{\pi}{2}, \dfrac{5\pi}{6}$	$x = 0, \dfrac{\pi}{2}, \pi$

23. $\qquad \cos 2x - \cos 3x = 0$

$-2 \sin \dfrac{5x}{2} \sin \left(\dfrac{-x}{2}\right) = 0$

$\sin \dfrac{5x}{2} = 0$	$\sin \dfrac{x}{2} = 0$
$\dfrac{5x}{2} = 0, \pi, 2\pi$	$\dfrac{x}{2} = 0$
$x = 0, \dfrac{2\pi}{5}, \dfrac{4\pi}{5}$	$x = 0$

25. $f(x) = \cos (3x+1) + \cos (3x-1)$

$\qquad = 2 \cos \dfrac{6x}{2} \cos \dfrac{2}{2}$

$\qquad = 2 \cos\ 3x \cos 1$

1. $1 + \sec x = \csc x (\tan x + \sin x)$

$$= \frac{1}{\sin x} \left(\frac{\sin x}{\cos x} + \sin x \right)$$

$$= \frac{1}{\cos x} + 1$$

3. $\sin x = \frac{1 - \cos^2 x}{\sin x}$

$$= \frac{\sin^2 x}{\sin x}$$

5. $\cos x + \sin x = \frac{2 \cos^2 x - 1}{\cos x - \sin x}$

$$= \frac{2 \cos^2 x - (\cos^2 x + \sin^2 x)}{\cos x - \sin x}$$

$$= \frac{2 \cos^2 x - \sin^2 x}{\cos x - \sin x}$$

$$= \frac{(\cos x + \sin x)(\cos x - \sin x)}{\cos x - \sin x}$$

7. $1 - \sin x = \frac{\cos x}{\sec x + \tan x}$

$$= \frac{\cos x}{\frac{1}{\cos x} + \frac{\sin x}{\cos x}}$$

$$= \frac{\cos^2 x}{1 + \sin x}$$

$$= \frac{1 - \sin^2 x}{1 + \sin x}$$

$$= \frac{(1 - \sin x)(1 + \sin x)}{1 + \sin x}$$

9. $\sec x - \cos x = \frac{\tan^2 x}{\sec x}$

$$= \frac{\sec^2 x - 1}{\sec x}$$

$$= \frac{\sec^2 x}{\sec x} - \frac{1}{\sec x}$$

13.

$$\sin x + 2 \cos^2 x = 1$$

$$\sin x + 2 (1 - \sin^2 x) = 1$$

$$\sin x + 2 - 2 \sin^2 x = 1$$

$$2 \sin^2 x - \sin x - 1 = 0$$

$$(2 \sin x + 1)(\sin x - 1) = 0$$

$$\sin x = -\frac{1}{2} \qquad \sin x = 1$$

$$x = \frac{7\pi}{6}, \frac{11\pi}{6} \qquad x = \frac{\pi}{2}$$

15.
$$\sec^5 x - 4 \sec x = 0$$

$$\sec x\,(\sec^2 x - 2)(\sec^2 x + 2) = 0$$

$\sec x = 0$	$\sec^2 x = 2$	$\sec^2 x = -2$
No solution	$\sec x = \sqrt{2}$	No solution

$$x = \frac{\pi}{4}, \frac{3\pi}{4}, \frac{5\pi}{4}, \frac{7\pi}{4}$$

17. $\cos A = \frac{3}{5}, \sin A = \frac{4}{5}$

$\sin B = \frac{2}{3}, \cos B = \dfrac{-\sqrt{5}}{3}$

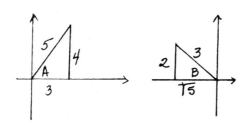

$$\sin (A+B) = \sin A \cos B + \cos A \sin B$$

$$= \frac{4}{5}\left(\frac{-\sqrt{5}}{3}\right) + \left(\frac{3}{5}\right)\left(\frac{2}{3}\right)$$

$$= \frac{-4\sqrt{5} + 6}{15}$$

19. $\tan 2A = \dfrac{2 \tan A}{1 - \tan^2 A}$

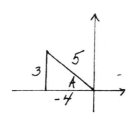

$$\cos A = \frac{-4}{5},\ \tan A = \frac{-3}{4}$$

$$\tan 2A = \frac{2\left(\frac{-3}{4}\right)}{1 - \left(\frac{-3}{4}\right)^2} = -\frac{24}{7}$$

21. $\cos\left(x - \frac{\pi}{2}\right) = \cos x \cos \frac{\pi}{6} + \sin x \sin \frac{\pi}{6}$

$$= \frac{\sqrt{3}}{2} \cos x + \frac{1}{2} \sin x$$

23. $\tan\left(2x + \frac{\pi}{4}\right) = \dfrac{\tan 2x + \tan \frac{\pi}{4}}{1 - \tan 2x \tan \frac{\pi}{4}} = \dfrac{\tan 2x + 1}{1 - \tan 2x}$

25. $\csc x = \frac{1}{2} \csc \frac{x}{2} \sec \frac{x}{2}$

$$= \frac{1}{2 \sin \frac{x}{2} \cos \frac{x}{2}}$$

$$= \frac{1}{\sin x}$$

27. $\cos(1+h) - \cos 1 = -2 \sin\left(\frac{2+h}{2}\right) \sin \frac{h}{2}$

29. $\dfrac{\cos 2x - \cos 4x}{\sin 2x + \sin 4x} = \dfrac{-2 \sin \frac{6x}{2} \sin\left(\frac{-2x}{2}\right)}{2 \sin \frac{6x}{2} \cos\left(\frac{-2x}{2}\right)} = \dfrac{-\sin(-x)}{\cos(-x)} = \dfrac{\sin x}{\cos x} = \tan x$

31. $\dfrac{\sin 3\theta}{\sin \theta} = \dfrac{\sin(2\theta + \theta)}{\sin \theta} = \dfrac{\sin 2\theta \cos \theta + \cos 2\theta \sin \theta}{\sin \theta} =$

$$= \dfrac{2 \sin \theta \cos^2 \theta + \cos 2\theta \sin \theta}{\sin \theta} = 2 \cos^2 \theta + \cos 2\theta$$

$$= 2(1 - \sin^2 \theta) + (1 - 2 \sin^2 \theta) = 3 - 4 \sin^2 \theta$$

33. $$\sin 4a \sin 2a + \sin^2 a = \frac{1}{2}(1 - \cos 6a)$$

$$\frac{1}{2}(\cos 2a - \cos 6a) + \sin^2 a =$$

$$\frac{1}{2}(1 - 2 \sin^2 a) - \frac{1}{2} \cos 6a + \sin^2 a = \frac{1}{2}(1 - \cos 6a)$$

35. $\cos 4x = 4 \cos 2x - 3 + 8 \sin^4 x$

$$= 4(1 - 2 \sin^2 x) - 3 + 8 \sin^4 x$$

$$= 1 - 8 \sin^2 x + 8 \sin^4 x$$

$$= 1 - 8 \sin^2 x (1 - \sin^2 x)$$

$$= 1 - 8 \sin^2 x \cos^2 x$$

$$= 1 - 2(2 \sin x \cos x)^2$$

$$= 1 - 2 \sin^2 2x$$

37.

$$\frac{1}{2}(\sigma_x - \sigma_z) = \left[\frac{1}{2}(\sigma_x + \sigma_z) + \tau_0 \cot \phi\right] \sin \phi$$

$$\frac{\sigma_x}{2} - \frac{\sigma_z}{2} = \frac{1}{2}(\sigma_x + \sigma_z) \sin \phi + \tau_0 \cot \phi \sin \phi$$

$$\frac{\sigma_x}{2} - \frac{\sigma_z}{2} = \frac{\sigma_x \sin \phi}{2} + \frac{\sigma_z \sin \phi}{2} + \tau_0 \cos \phi$$

$$\frac{\sigma_x}{2}(1 - \sin \phi) = \frac{\sigma_z}{2}(1 + \sin \phi) + \tau_0 \cos \phi$$

$$\sigma_x = \sigma_z \frac{(1 + \sin \phi)}{(1 - \sin \phi)} + 2\tau_0 \frac{\cos \phi}{1 - \sin \phi}$$

$$= \sigma_z\, b + 2\, \tau_0 \frac{\cos \phi}{1 - \sin \phi}$$

$$b = \frac{1 + \sin \phi}{1 - \sin \phi} \cdot \frac{1 - \sin \phi}{1 - \sin \phi} = \frac{1 - \sin^2 \phi}{(1 - \sin \phi)^2} = \frac{\cos^2 \phi}{(1 - \sin \phi)^2}$$

$$\sqrt{b} = \frac{\cos \phi}{1 - \sin \phi}$$

39. $\tan 2\theta = \frac{-4}{3}$, $\cos 2\theta = \frac{-3}{5}$

$$\sin \theta = \sqrt{\frac{1 - \cos 2\theta}{2}} = \sqrt{\frac{1 + 3/5}{2}} = \frac{2}{\sqrt{5}}$$

$$\cos \theta = \sqrt{\frac{1 + \cos 2\theta}{2}} = \sqrt{\frac{1 - 3/5}{2}} = \frac{1}{\sqrt{5}}$$

41. $\sin x + \sin 2x = 2 \sin \frac{3x}{2} \cos\left(\frac{-x}{2}\right) = 2 \sin \frac{3x}{2} \cos \frac{x}{2}$

Chapter 10   Systems of Equations and Inequalities

Section 10.1   Linear Systems of Equations

1. $2x - 4y = 2$
   $-2x + y = 4$

   $\qquad -3y = 6, \quad y = -2$

   $-2x - 2 = 4, \quad x = -3$

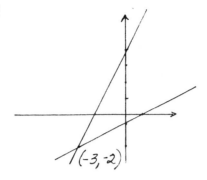

$(-3, -2)$

3. $3x + 4y = \quad 23$
   $x - 3y = -1$

   $3x + 4y = \quad 23$
   $-3x + 9y = + 3$

   $\qquad 13y = \quad 26, \quad y = 2$

   $x - 6 = -1, \quad x = 5$

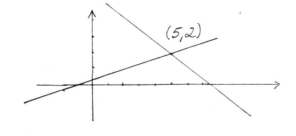

$(5, 2)$

5. $x - 4y = \quad 8$
   $2x - 8y = \quad 16$

   $-2x + 8y = -16$
   $2x - 8y = \quad 16$

   $\qquad 0 = \quad 0, \quad \text{dependent}$

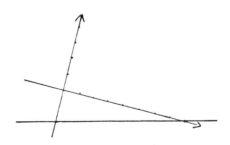

7. $3x + 4y = 10$
   $6x + 8y = -2$

   $-6x - 8y = -20$
   $6x + 8y = -2$

   $0 = -22,$ inconsistent

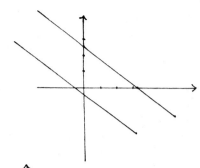

9. $3x - y = 2$
   $2x + y = 6$
   $5x = 8,$ $x = \frac{8}{5}$

   $2\left(\frac{8}{5}\right) + y = 6,$ $y = \frac{14}{5}$

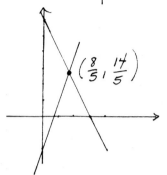

$\left(\frac{8}{5}, \frac{14}{5}\right)$

11. $3x + y = 0$
    $2x - 2y = 2$

    $6x + 2y = 0$
    $2x - 2y = 2$
    $8x = 2,$ $x = \frac{1}{4}$

    $3\left(\frac{1}{4}\right) + y = 0,$ $y = \frac{-3}{4}$

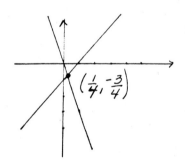

$\left(\frac{1}{4}, \frac{-3}{4}\right)$

13. $x + 7y - 7 = 0$
    $8x - 7y - 3 = 0$
    $9x = 10,$ $x = \frac{10}{9}$

    $\frac{10}{9} + 7y - 7 = 0,$ $y = \frac{53}{63}$

$\left(\frac{10}{9}, \frac{53}{63}\right)$

15. $r - s = 3$
    $\underline{r - 2s = 5}$
    $r = 3 + s$
    $(3+s) - 2s = 5, \quad s = -2$
    $r = 3-2 = 1$

17. $x + 3y = 7$
    $\underline{2x + 6y = 9}$
    $x = 7 - 3y$
    $2(7-3y) + 6y = 9$
    $14 = 9$, inconsistent

19. $5y - x = 2$
    $\underline{2y + 3x = 11}$
    $x = 5y-2,$
    $2y + 3(5y-2) = 11, \ y=1$
    $x = 5(1) -2 = 3$

21. $4x + 5y = -6$
    $\underline{3x - 2y = -16}$
    $x = \dfrac{-6 - 5y}{4}$
    $3\left(\dfrac{-6-5y}{4}\right) - 2y = -16,$
    $-18-15y-2y(4) = 64, \ y=2$
    $x = \dfrac{-6-5(2)}{4} = -4$

23. $6x - 8y = 14$
    $\underline{3x - 4y = 7}$
    $x = \dfrac{7 + 4y}{3}$
    $6\left(\dfrac{7+4y}{3}\right) - 8y = 14$
    $14 + 8y - 8y = 14$
    dependent

25. $3x + 5y = 5$
    $\underline{x + 4y = 11}$
    $x = 11 - 4y$
    $3(11-4y) + 5y = 5, \ y=4$
    $x = 11 - 4(4) = -5$

27. $9z - 13w = -3$
    $\underline{6z - 7w = 3}$
    $z = \dfrac{7w + 3}{6}$
    $9\left(\dfrac{7w+3}{6}\right) - 13w = -3$
    $9 + 21w - 26w = -6, \ w=3$
    $z = \dfrac{7(3) + 3}{6} = 4$

29. $2x + 3y = -10$
    $\underline{3x - 2y = -2}$
    $x = \dfrac{-10 - 3y}{2}$
    $3\left(\dfrac{-10 - 3y}{2}\right) - 2y = -2$
    $-30 - 9y - 4y = -4, \ y = +-2$
    $x = \dfrac{-10 - 3(-2)}{2} = -2$

31. $A + B = 25$
    $\underline{A - B = 3}$
    $2A = 28, \quad A = 14$
    $14 + B = 25, \quad B = 18$

33. $x + y = 12$

$\underline{x - y = 4}$

$2x = 16, \quad x = 8$

$ 8 + y = 12, \quad y = 4$

35. $ax + by^2 = 2$

$(4, -2) \rightarrow 4a + 4b = 2$

$(-2, 2) \rightarrow \underline{-2a + 4b = 2}$

$ 6a = 0, \quad a = 0$

$ 4b = 2, \quad b = \frac{1}{2}$

37. rate $\times$ time = distance

$(v - w)\ 10 = 60$

$\underline{(v + w)\ 8 = 60}$

$80v - 80w = 480$

$\underline{80v + 80w = 600}$

$ 160v = 1080$

$ v = 6.75$

$67.5 - 10w = 60, \quad w = .75$

39.

	Lead	Zinc
A_1	.80	.20
A_2	.50	.50
$A_1 + A_2$	.60	.40

$.80A_1 + .50A_2 = .60\ (10)$

$.20A_1 + .50A_2 = .40\ (10)$

$.60A_1 = 2, \quad A_1 = \frac{10}{3}$

$A_1 + A_2 = 10, \quad A_2 = 10 - \frac{10}{3} = \frac{20}{3}$

41. $n + d = 35$

$\underline{.05n + .10d = 2.50}$

$10n + 10d = 350$

$\underline{5n + 10d = 250}$

$ 5n = 100, \quad n = 20$

$ d = 35 - n$

$ = 35 - 20$

$ = 15$

43. $C_1 = x$ min to do job

$C_2 = y$ min to do job

$\frac{1}{x}$ = amt done by C_1 in 1 min

$\frac{1}{y}$ = amt done by C_2 in 1 min

$10(\frac{1}{x} + \frac{1}{y}) = 1$

$6(\frac{1}{x} + \frac{1}{y}) + 9(\frac{1}{y}) = 1$

$\frac{10}{x} + \frac{10}{y} = 1$

$\underline{\frac{6}{x} + \frac{15}{y} = 1}$

$\frac{60}{x} + \frac{60}{y} = 6$

$\underline{-\frac{60}{x} - \frac{150}{y} = -10}$

$ -\frac{90}{y} = -4, \quad y = 22.5 \text{ min}$

$\frac{10}{x} + \frac{10}{22.5} = 1$

$10(22.5) + 10x = 22.5$

$x = 18 \text{ min}$

45. $\dfrac{x+550}{y} = \tan 29.1°$

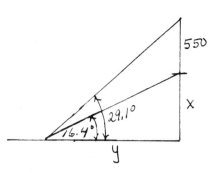

$\dfrac{x}{y} = \tan 16.4°,\ x = y \tan 16.4°$

$y \tan 16.4° + 550 = y \tan 29.1°$

$y = \dfrac{-550}{\tan 16.4° - \tan 29.1°} = 2097$

$x = 2097 \tan 16.4° = 617.2$

47. Demand: $\begin{array}{ll} x = 100 & p = 400 \\ x = 150 & p = 250 \end{array}$

$p - 400 = \left(\dfrac{400-250}{100-150}\right)(x-100)$

$p = -3x + 700$

Supply: $\begin{array}{ll} x = 75 & p = 300 \\ x = 150 & p = 350 \end{array}$

$p - 300 = \left(\dfrac{350-300}{150-75}\right)(x-75)$

$p = \dfrac{2x}{3} + 250$

$-3x + 700 = \dfrac{2x}{3} + 250$

$x = 123,\ p = -3(123) + 700 = 331$

Section 10.2   Nonlinear Systems

1.

3.

5.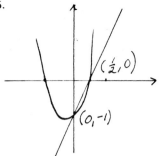

$(\frac{1}{2}, 0)$

$(0, -1)$

7.

9.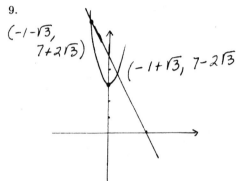

$(-1-\sqrt{3}, \; 7+2\sqrt{3})$

$(-1+\sqrt{3}, \; 7-2\sqrt{3})$

11.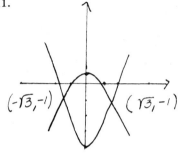

$(-\sqrt{3}, -1)$ $(\sqrt{3}, -1)$

13.
$$\begin{array}{rcl} x^2 + y^2 &=& 4 \\ 4x^2 + y^2 &=& 4 \\ \hline -3x^2 &=& 0 \\ x &=& 0 \end{array}$$

$$y^2 = 4, \quad y = \pm 2$$

$$(0, 2), \quad (0, -2)$$

15.
$$\begin{array}{rcl} x^2 + 4y^2 &=& 2 \\ x^2 - y^2 &=& 3 \\ \hline 5y^2 &=& -1 \end{array}$$

No solution

17.
$$\begin{array}{rcl} x^2 + y^2 &=& 1 \\ 2x^2 - y^2 &=& 2 \end{array}$$

$$3x^2 = 3$$

$$x^2 = 1, \quad x = \pm 1$$

$$y^2 = 0,$$

$$(1, 0), \quad (-1, 0)$$

19.
$$\begin{array}{rcl} y - x &=& 1 \\ x^2 - y^2 &=& 10 \end{array}$$

$$y = 1 + x$$

$$x^2 - (1+x)^2 = 10$$

$$x^2 - 1 - 2x - x^2 = 10$$

$$x = \frac{-11}{2}$$

$$y = 1 - \frac{11}{2} = \frac{-9}{2}$$

$$\left(-\frac{11}{2}, \frac{-9}{2}\right)$$

21. $y^2 + 2x^2 = 2$
    $\underline{y^2 + x^2 = 2}$

    $x^2 = 0, \quad x = 0$
    $y^2 = 2, \quad y = \pm\sqrt{2}$

    $(0, \sqrt{2}), \quad (0, -\sqrt{2})$

23. $2a^2 - 4b^2 = 7$
    $\underline{a^2 + 2b^2 = 3}$

    $2a^2 - 4b^2 = 7$
    $\underline{2a^2 - 4b^2 = 6}$

    $4a^2 = 13, \quad a^2 = \dfrac{13}{4}$

    $2b^2 = 3 - a^2 = 3 - \dfrac{13}{4}$

    $b^2 = \dfrac{-1}{8}$, no solution

25. $ab = 16$
    $\underline{a - b = 0}$

    $a = b$
    $b^2 = 16, \quad b = \pm 4$

    $a = \pm 4$

    $(4, 4), \quad (-4, -4)$

27. $xy = 1$
    $\underline{x^2 + y^2 = 1}$

    $x = \dfrac{1}{y}$

    $\dfrac{1}{y^2} + y^2 = 1$

    $1 + y^4 = y^2$

    $y^4 - y^2 + 1 = 0$

    $y^4 = \dfrac{1 \pm \sqrt{1-4}}{2}$

    No real solutions

29. $x^2 + y^2 = 16$
    $y = 2$

    $x^2 + 4 = 16$
    $x^2 = 12, \quad x = \pm\sqrt{12}$
    $(2\sqrt{3}, 2), \quad (-2\sqrt{3}, 2)$

31. $x + y = 1$
    $\underline{x^2 - y^2 = 1}$

    $x = 1 - y$
    $(1-y)^2 - y^2 = 1$
    $1 - 2y + y^2 - y^2 = 1$
    $y = 0$
    $x = 1, \quad (1, 0)$

33. $x + y = 30$
    $\underline{xy = 225}$

    $x = 30 - y$

    $(30-y)y = 225$
    $30y - y^2 - 225 = 0$
    $(y-15)(y-15) = 0$
    $y = 15, \ x = 15$

35. $p = x + 6$
    $p = \dfrac{208}{x + 3}$

    $x + 6 = \dfrac{208}{x + 3}$

    $(x+6)(x+3) = 208$
    $x^2 + 9x + 18 - 208 = 0$
    $x^2 + 9x - 190 = 0$
    $(x-10)(x+19) = 0$
    $x = 10, \ \cancel{-19}$
    $p = 10 + 6 = 16$

1. $\begin{bmatrix} 1 & 2 & | & 10 \\ 1 & -2 & | & 6 \end{bmatrix} \xrightarrow{-①+②} \begin{bmatrix} 1 & 2 & | & 10 \\ 0 & -4 & | & -16 \end{bmatrix} \xrightarrow{-\frac{1}{4}②} \begin{bmatrix} 1 & 2 & | & 10 \\ 0 & 1 & | & 4 \end{bmatrix}$

$x + 2y = 10$

$y = 4,\ x = 10 - 2(4) = 2$

3. $\begin{bmatrix} 3 & -1 & | & 5 \\ 1 & 4 & | & 6 \end{bmatrix} \xrightarrow{\frac{1}{3}①} \begin{bmatrix} 1 & -\frac{1}{3} & | & \frac{5}{3} \\ 1 & 4 & | & 6 \end{bmatrix} \xrightarrow{-①+②} \begin{bmatrix} 1 & -\frac{1}{3} & | & \frac{5}{3} \\ 0 & \frac{13}{3} & | & \frac{13}{3} \end{bmatrix}$

$x - \frac{1}{3}y = \frac{5}{3}$

$y = 1,\ x = \frac{5}{3} + \frac{1}{3} = 2$

5. $\begin{bmatrix} 8 & 5 & 0 & | & 4 \\ 0 & 5 & -3 & | & 1 \\ 12 & 0 & 5 & | & 6 \end{bmatrix} \xrightarrow{\frac{1}{8}①} \begin{bmatrix} 1 & \frac{5}{8} & 0 & | & \frac{1}{2} \\ 0 & 5 & -3 & | & 1 \\ 12 & 0 & 5 & | & 6 \end{bmatrix} \xrightarrow{-12①+③}$

$\begin{bmatrix} 1 & \frac{5}{8} & 0 & | & \frac{1}{2} \\ 0 & 5 & -3 & | & 1 \\ 0 & -\frac{15}{2} & 5 & | & 0 \end{bmatrix} \xrightarrow{\frac{1}{5}②} \begin{bmatrix} 1 & \frac{5}{8} & 0 & | & \frac{1}{2} \\ 0 & 1 & -\frac{3}{5} & | & \frac{1}{5} \\ 0 & -\frac{15}{2} & 5 & | & 0 \end{bmatrix} \xrightarrow[-\frac{5}{8}②+①]{\frac{15}{2}②+③}$

$\begin{bmatrix} 1 & 0 & \frac{3}{8} & | & \frac{3}{8} \\ 0 & 1 & -\frac{3}{5} & | & \frac{1}{5} \\ 0 & 0 & \frac{1}{2} & | & \frac{3}{2} \end{bmatrix}$

$x + \frac{3}{8}z = \frac{3}{8}$

$y - \frac{3z}{5} = \frac{1}{5}$

$\frac{1}{2}z = \frac{3}{2},\ z = 3$

$y = \frac{1}{5} + \frac{3}{5}(3) = 2$

$x = \frac{3}{8} - \frac{3}{8}(3) = -\frac{3}{4}$

7.

$$\left[\begin{array}{ccc|c} 3 & 5 & -3 & 31 \\ 2 & -3 & 2 & 13 \\ 5 & 2 & -5 & 20 \end{array}\right] \xrightarrow{-①+③} \left[\begin{array}{ccc|c} 3 & 5 & --3 & 31 \\ 2 & -3 & 2 & 13 \\ 2 & -3 & -2 & -11 \end{array}\right] \begin{array}{l} -②+① \\ \hline \\ -②+③ \end{array}\rightarrow$$

$$\left[\begin{array}{ccc|c} 1 & 8 & -5 & 18 \\ 2 & -3 & 2 & 13 \\ 0 & 0 & -4 & -24 \end{array}\right] \begin{array}{l} -\frac{1}{4}③ \\ \hline \\ -2①+② \end{array}\rightarrow \left[\begin{array}{ccc|c} 1 & 8 & -5 & 18 \\ 0 & -19 & 12 & -23 \\ 0 & 0 & 1 & 6 \end{array}\right]$$

$$x + 8y - 5z = 18$$
$$-19y + 12z = -23$$
$$z = 6$$
$$-19y = -23 - 12(6), \quad y = 5$$
$$x = -8(5) + 5(6) + 18 = 8$$

9.

$$\left[\begin{array}{ccc|c} 2 & 0 & -5 & 2 \\ 0 & 9 & -4 & 7 \\ 3 & -12 & 0 & -2 \end{array}\right] \xrightarrow{-①+③} \left[\begin{array}{ccc|c} 2 & 0 & -5 & 2 \\ 0 & 9 & -4 & 7 \\ 1 & -12 & 5 & -4 \end{array}\right] \xrightarrow{-③+①}$$

$$\left[\begin{array}{ccc|c} 1 & 12 & -10 & 6 \\ 0 & 9 & -4 & 7 \\ 1 & -12 & 5 & 4 \end{array}\right] \xrightarrow{-①+③} \left[\begin{array}{ccc|c} 1 & 12 & -10 & 6 \\ 0 & 9 & -4 & 7 \\ 0 & -24 & 15 & -10 \end{array}\right] \xrightarrow{\frac{1}{9}②}$$

$$\left[\begin{array}{ccc|c} 1 & 12 & -10 & 6 \\ 0 & 1 & -\frac{4}{9} & \frac{7}{9} \\ 0 & -24 & 15 & -10 \end{array}\right] \xrightarrow{24②+③} \left[\begin{array}{ccc|c} 1 & 12 & -10 & 6 \\ 0 & 1 & -\frac{4}{9} & \frac{7}{9} \\ 0 & 0 & \frac{13}{3} & \frac{26}{3} \end{array}\right]$$

$$x + 12y - 10z = 6$$
$$y - \frac{4}{9}z = \frac{7}{9}$$
$$\frac{13z}{3} = \frac{26}{3}, \ z = 2$$
$$y = \frac{7}{9} + \frac{4}{9}(2) = \frac{5}{3}$$
$$x = 6 - 12\left(\frac{5}{3}\right) + 10(2) = 6$$

11.
$$\begin{bmatrix} 1 & 1 & 2 & | & 3 \\ 3 & -1 & -1 & | & 1 \\ 1 & 5 & 9 & | & 11 \end{bmatrix} \xrightarrow[-①+③]{-3①+②} \begin{bmatrix} 1 & 1 & 2 & | & 3 \\ 0 & -4 & -7 & | & -8 \\ 0 & 4 & 7 & | & 8 \end{bmatrix} \xrightarrow{②+③}$$

$$\begin{bmatrix} 1 & 1 & 2 & | & 3 \\ 0 & -4 & -7 & | & 8 \\ 0 & 0 & 0 & | & 0 \end{bmatrix} \quad \text{Dependent}$$

13.
$$\begin{bmatrix} 2 & 1 & -1 & | & 9 \\ 1 & -1 & 1 & | & 0 \\ -1 & 3 & -2 & | & 5 \end{bmatrix} \xrightarrow{③+①} \begin{bmatrix} 1 & 4 & 3 & | & 14 \\ 1 & -1 & 1 & | & 0 \\ -1 & 3 & -2 & | & 5 \end{bmatrix} \xrightarrow{②+③}$$

$$\begin{bmatrix} 1 & 4 & -3 & | & 14 \\ 1 & -1 & 1 & | & 0 \\ 0 & 2 & -1 & | & 5 \end{bmatrix} \xrightarrow{-①+②} \begin{bmatrix} 1 & 4 & -3 & | & 14 \\ 0 & -5 & 4 & | & -14 \\ 0 & 2 & -1 & | & 5 \end{bmatrix} \xrightarrow[\frac{1}{2}③]{-\frac{1}{5}②}$$

$$\begin{bmatrix} 1 & 4 & -3 & | & 14 \\ 0 & 1 & -\frac{4}{5} & | & \frac{14}{5} \\ 0 & 1 & -\frac{1}{2} & | & \frac{5}{2} \end{bmatrix} \xrightarrow{-②+③} \begin{bmatrix} 1 & 4 & -3 & | & 14 \\ 0 & 1 & -\frac{4}{5} & | & \frac{14}{5} \\ 0 & 0 & \frac{3}{10} & | & -\frac{3}{10} \end{bmatrix}$$

$$x + 4y - 3z = 14$$
$$y - \frac{4}{5}z = \frac{14}{5}$$
$$\frac{3z}{10} = \frac{-3}{10}, \; z = -1$$
$$y = \frac{14}{5} + \frac{4}{5}(-1) = 2$$
$$x = 14 - 4(2) + 3(-1) = 3$$

15.
$$\begin{bmatrix} 1 & -1 & 0 & | & 2 \\ 0 & 1 & -1 & | & -3 \\ 3 & 2 & -5 & | & 1 \end{bmatrix} \xrightarrow{-3①+③} \begin{bmatrix} 1 & -1 & 0 & | & 2 \\ 0 & 1 & -1 & | & -3 \\ 0 & 5 & -5 & | & -5 \end{bmatrix} \xrightarrow{-5②+③}$$

$$\begin{bmatrix} 1 & -1 & 0 & | & 2 \\ 0 & 1 & -1 & | & -3 \\ 0 & 0 & 0 & | & 10 \end{bmatrix} \quad \text{Inconsistent}$$

- 150 -

17.

$$\begin{bmatrix} 1 & -3 & 2 & | & 0 \\ 2 & 1 & -1 & | & -3 \\ 10 & -2 & 0 & | & -12 \end{bmatrix} \xrightarrow{\substack{-2①+② \\ -10①+③}} \begin{bmatrix} 1 & -3 & 2 & | & 0 \\ 0 & 7 & -5 & | & -3 \\ 0 & 28 & -20 & | & -12 \end{bmatrix} \xrightarrow{-4②+③}$$

$$\begin{bmatrix} 1 & -3 & 2 & | & 0 \\ 0 & 7 & 5 & | & -3 \\ 0 & 0 & 0 & | & 0 \end{bmatrix} \qquad \text{Dependent}$$

19.

$$\begin{bmatrix} 1 & 1 & 1 & | & 3 \\ 1 & -2 & -3 & | & 4 \\ 2 & 1 & 2 & | & 9 \end{bmatrix} \xrightarrow{\substack{-①+② \\ -2①+③}} \begin{bmatrix} 1 & 1 & 1 & | & 3 \\ 0 & -3 & -4 & | & 1 \\ 0 & -1 & 0 & | & 3 \end{bmatrix}$$

$$x + y + z = 3$$
$$-3y - 4z = 1$$
$$-y = 3, \quad y = -3$$
$$4z = -1 - 3(-3), \; z = 2$$
$$x = 3 - (-3) - (2) = 4$$

21.

$$\begin{bmatrix} 3 & -2 & 1 & 1 & | & 10 \\ 1 & 1 & -1 & -1 & | & -5 \\ 2 & 1 & 1 & 5 & | & 18 \\ -1 & -2 & 5 & -2 & | & 5 \end{bmatrix} \xrightarrow{\substack{④+② \\ 2④+③ \\ 3④+①}} \begin{bmatrix} 0 & -8 & 16 & -5 & | & 25 \\ 0 & -1 & 4 & -3 & | & 0 \\ 0 & -3 & 11 & 1 & | & 28 \\ -1 & -2 & 5 & -2 & | & 5 \end{bmatrix} \xrightarrow{\substack{-④ \\ ④\leftrightarrow① \\ -②}}$$

$$\begin{bmatrix} 1 & 2 & -5 & 2 & | & -5 \\ 0 & 1 & -4 & 3 & | & 0 \\ 0 & -3 & 11 & 1 & | & 28 \\ 0 & -8 & 16 & -5 & | & 25 \end{bmatrix} \xrightarrow{\substack{-2②+① \\ 3②+③ \\ 8②+④}} \begin{bmatrix} 1 & 0 & 3 & -4 & | & -5 \\ 0 & 1 & -4 & 3 & | & 0 \\ 0 & 0 & -1 & 10 & | & 28 \\ 0 & 0 & -16 & 19 & | & 25 \end{bmatrix} \xrightarrow{\substack{3③+① \\ -4③+② \\ -16③+④}}$$

$$\begin{bmatrix} 1 & 0 & 0 & 26 & | & 79 \\ 0 & 1 & 0 & -37 & | & -112 \\ 0 & 0 & -1 & 10 & | & 28 \\ 0 & 0 & 0 & -141 & | & -423 \end{bmatrix}$$

$$x + 26w = 79$$
$$y - 37w = -112$$
$$-z + 10w = 28$$
$$-141w = -423, \; w = 3$$
$$z = -28 + 10(3) = 2$$
$$y = -112 + 37(3) = -1$$
$$x = 79 - 26(3) = 1$$

23.

$$\left[\begin{array}{cccc|c} 3 & 1 & 1 & -1 & -2 \\ -5 & 2 & -3 & -3 & -1 \\ 7 & 5 & 4 & -1 & -1 \\ 4 & -1 & -2 & 3 & 7 \end{array}\right] \xrightarrow{①+④} \left[\begin{array}{cccc|c} 3 & 1 & 1 & -1 & -2 \\ -5 & 2 & -3 & -3 & -1 \\ 7 & 5 & 4 & -1 & -1 \\ 7 & 0 & -1 & 2 & 5 \end{array}\right] \xrightarrow[2①+②]{-④+③}$$

$$\left[\begin{array}{cccc|c} 3 & 1 & 1 & -1 & -2 \\ 1 & 4 & -1 & -5 & -5 \\ 0 & 5 & 5 & -3 & -6 \\ 7 & 0 & -1 & 2 & 5 \end{array}\right] \xrightarrow[3②+①]{-7②+④} \left[\begin{array}{cccc|c} 0 & -11 & 4 & 14 & 13 \\ 1 & 4 & -1 & -5 & -5 \\ 0 & 5 & 5 & -3 & -6 \\ 0 & -28 & 6 & 37 & 40 \end{array}\right] \xrightarrow{-3①+④}$$

$$\left[\begin{array}{cccc|c} 0 & -11 & 4 & 14 & 13 \\ 1 & 4 & -1 & -5 & -5 \\ 0 & 5 & 5 & -3 & -6 \\ 0 & 5 & -6 & -5 & 1 \end{array}\right] \xrightarrow[2③+①]{-③+④} \left[\begin{array}{cccc|c} 0 & -1 & 14 & 8 & 1 \\ 1 & 4 & -1 & -5 & -5 \\ 0 & 5 & 5 & -3 & -6 \\ 0 & 0 & -11 & -2 & 7 \end{array}\right] \xrightarrow{5①+③}$$

$$\left[\begin{array}{cccc|c} 0 & -1 & 14 & 8 & 1 \\ 1 & 4 & -1 & -5 & -5 \\ 0 & 0 & 75 & 37 & -1 \\ 0 & 0 & -11 & -2 & 7 \end{array}\right] \xrightarrow[-①]{7④+③} \left[\begin{array}{cccc|c} 0 & 1 & -14 & -8 & -1 \\ 1 & 4 & -1 & -5 & -5 \\ 0 & 0 & -2 & 23 & 48 \\ 0 & 0 & -11 & -2 & 7 \end{array}\right] \xrightarrow[①\leftrightarrow②]{5③+④}$$

$$\left[\begin{array}{cccc|c} 1 & 4 & -1 & -5 & -5 \\ 0 & 1 & -14 & -8 & -1 \\ 0 & 0 & -2 & 23 & 48 \\ 0 & 0 & -1 & -117 & -233 \end{array}\right] \xrightarrow{-2④+③} \left[\begin{array}{cccc|c} 1 & 4 & -1 & -5 & -5 \\ 0 & 1 & -14 & -8 & -1 \\ 0 & 0 & 0 & 257 & 514 \\ 0 & 0 & -1 & -117 & -233 \end{array}\right]$$

$$x + 4y - z - 5w = -5$$
$$y - 14z - 8w = -1$$
$$257w = 514$$
$$-z - 117w = -233$$

$$w = 2$$
$$z = 233 - 117(2) = -1$$
$$y = -1 + 8(2) + 14(-1) = 1$$
$$x = -5 + 5(2) - (-1) - 4(1) = 1$$

25. $n + d + q = 900$

$n = 20 + d, \qquad n - d = 20$

$n + d = 44 + q, \qquad n + d - q = 44$

$$\left[\begin{array}{ccc|c} 1 & 1 & 1 & 900 \\ 1 & -1 & 0 & 20 \\ 1 & 1 & -1 & 44 \end{array}\right] \begin{array}{c} -①+② \\ \hline -①+③ \end{array} \left[\begin{array}{ccc|c} 1 & 1 & 1 & 900 \\ 0 & -2 & -1 & -880 \\ 0 & 0 & -2 & -856 \end{array}\right] \begin{array}{c} -\frac{1}{2}③ \\ \hline -② \end{array}$$

$$\left[\begin{array}{ccc|c} 1 & 1 & 1 & 900 \\ 0 & 2 & 1 & 880 \\ 0 & 0 & 1 & 428 \end{array}\right] \begin{array}{c} -③+② \\ \hline -③+① \end{array} \left[\begin{array}{ccc|c} 1 & 1 & 0 & 472 \\ 0 & 2 & 0 & 452 \\ 0 & 0 & 1 & 428 \end{array}\right] \begin{array}{c} \frac{1}{2}② \end{array}$$

$$\left[\begin{array}{ccc|c} 1 & 1 & 0 & 472 \\ 0 & 1 & 0 & 226 \\ 0 & 0 & 1 & 428 \end{array}\right] \begin{array}{c} -②+① \end{array} \left[\begin{array}{ccc|c} 1 & 0 & 0 & 246 \\ 0 & 1 & 0 & 226 \\ 0 & 0 & 1 & 428 \end{array}\right]$$

27. $l_1 + l_2 + l_3 = 45$

$2l_1 = l_2 + l_3, \qquad 2l_1 - l_2 - l_3 = 0$

$l_2 = 4 + l_3, \qquad l_2 - l_3 = 4$

$$\left[\begin{array}{ccc|c} 1 & 1 & 1 & 45 \\ 2 & -1 & -1 & 0 \\ 0 & 1 & -1 & 4 \end{array}\right] \xrightarrow{-2①+②} \left[\begin{array}{ccc|c} 1 & 1 & 1 & 45 \\ 0 & -3 & -3 & -90 \\ 0 & 1 & -1 & 4 \end{array}\right] \xrightarrow{-\frac{1}{3}②}$$

$$\left[\begin{array}{ccc|c} 1 & 1 & 1 & 45 \\ 0 & 1 & 1 & 30 \\ 0 & 1 & -1 & 4 \end{array}\right] \xrightarrow{-②+③} \left[\begin{array}{ccc|c} 1 & 1 & 1 & 45 \\ 0 & 1 & 1 & 30 \\ 0 & 0 & -2 & -26 \end{array}\right] \xrightarrow{-\frac{1}{2}③}$$

$$\left[\begin{array}{ccc|c} 1 & 1 & 1 & 45 \\ 0 & 1 & 1 & 30 \\ 0 & 0 & 1 & 13 \end{array}\right] \begin{array}{c} -③+② \\ \hline -③+① \end{array}\!\!\longrightarrow \left[\begin{array}{ccc|c} 1 & 1 & 0 & 32 \\ 0 & 1 & 0 & 17 \\ 0 & 0 & 1 & 13 \end{array}\right] \xrightarrow{-②+①}$$

$$\left[\begin{array}{ccc|c} 1 & 0 & 0 & 15 \\ 0 & 1 & 0 & 17 \\ 0 & 0 & 1 & 13 \end{array}\right]$$

1. $A + B = \begin{bmatrix} 5 & 0 \\ 1 & 1 \end{bmatrix} + \begin{bmatrix} -1 & 2 \\ 1 & 0 \end{bmatrix} = \begin{bmatrix} 4 & 2 \\ 2 & 1 \end{bmatrix}$

$A + B = \begin{bmatrix} 5 & 0 \\ 1 & 1 \end{bmatrix} - \begin{bmatrix} -1 & 2 \\ 1 & 0 \end{bmatrix} = \begin{bmatrix} 6 & -2 \\ 0 & 1 \end{bmatrix}$

$2A = 2\begin{bmatrix} 5 & 0 \\ 1 & 1 \end{bmatrix} = \begin{bmatrix} 10 & 0 \\ 2 & 2 \end{bmatrix}$

3. $A + B = \begin{bmatrix} 1 & 2 & 3 \\ 4 & 0 & 1 \end{bmatrix} + \begin{bmatrix} 1 & 0 & 2 \\ 0 & 5 & 6 \end{bmatrix} = \begin{bmatrix} 2 & 2 & 5 \\ 4 & 5 & 7 \end{bmatrix}$

$A - B = \begin{bmatrix} 1 & 2 & 3 \\ 4 & 0 & 1 \end{bmatrix} - \begin{bmatrix} 1 & 0 & 2 \\ 0 & 5 & 6 \end{bmatrix} = \begin{bmatrix} 0 & 2 & 1 \\ 4 & -5 & -5 \end{bmatrix}$

$2A = 2\begin{bmatrix} 1 & 2 & 3 \\ 4 & 0 & 1 \end{bmatrix} = \begin{bmatrix} 2 & 4 & 6 \\ 8 & 0 & 2 \end{bmatrix}$

5. $A + B = \begin{bmatrix} 1 \\ 0 \\ 2 \end{bmatrix} + \begin{bmatrix} 7 \\ -1 \\ 5 \end{bmatrix} = \begin{bmatrix} 8 \\ -1 \\ 7 \end{bmatrix}$

$A - B = \begin{bmatrix} 1 \\ 0 \\ 2 \end{bmatrix} + \begin{bmatrix} 7 \\ -1 \\ 5 \end{bmatrix} = \begin{bmatrix} -6 \\ 1 \\ -3 \end{bmatrix}$

$2A = 2\begin{bmatrix} 1 \\ 0 \\ 2 \end{bmatrix} = \begin{bmatrix} 2 \\ 0 \\ 4 \end{bmatrix}$

7. $A + B = \begin{bmatrix} 0 & 3 & 1 \end{bmatrix} + \begin{bmatrix} -1 & 2 & 0 \end{bmatrix} = \begin{bmatrix} -1 & 5 & 1 \end{bmatrix}$

$\quad A - B = \begin{bmatrix} 0 & 3 & 1 \end{bmatrix} - \begin{bmatrix} -1 & 2 & 0 \end{bmatrix} = \begin{bmatrix} 1 & 1 & 1 \end{bmatrix}$

$\quad\quad 2A = 2\begin{bmatrix} 0 & 3 & 1 \end{bmatrix} - \begin{bmatrix} 0 & 6 & 2 \end{bmatrix}$

11. $AB = \begin{bmatrix} 5 & 0 \\ 1 & 1 \end{bmatrix}\begin{bmatrix} -1 & 2 \\ 1 & 0 \end{bmatrix} = \begin{bmatrix} 5(-1)+0(1) & 5(2)+0(0) \\ 1(-1)+1(1) & 1(2)+1(0) \end{bmatrix} = \begin{bmatrix} -5 & 10 \\ 0 & 2 \end{bmatrix}$

$\quad BA = \begin{bmatrix} -1 & 2 \\ 1 & 0 \end{bmatrix}\begin{bmatrix} -5 & 0 \\ 1 & 1 \end{bmatrix} = \begin{bmatrix} -1(5)+2(1) & -1(0)+2(1) \\ 1(5)+0(1) & 1(0)+0(1) \end{bmatrix} = \begin{bmatrix} -3 & 2 \\ 5 & 0 \end{bmatrix}$

15. $AB = \begin{bmatrix} -1 \\ 5 \end{bmatrix}\begin{bmatrix} 2 & 0 \end{bmatrix} = \begin{bmatrix} -2 & 0 \\ 10 & 0 \end{bmatrix}$

$\quad BA = \begin{bmatrix} 2 & 0 \end{bmatrix}\begin{bmatrix} -1 \\ 5 \end{bmatrix} = \begin{bmatrix} 2(-1)+0(5) \end{bmatrix} = \begin{bmatrix} -2 \end{bmatrix}$

17. $AB = \begin{bmatrix} 2 & 1 \\ 1 & -1 \end{bmatrix}\begin{bmatrix} 1 \\ 2 \end{bmatrix} = \begin{bmatrix} 2(-1)+1(2) \\ 1(1)-1(2) \end{bmatrix} = \begin{bmatrix} 4 \\ -1 \end{bmatrix}$

$\quad BA$ not possible

19. $AB = \begin{bmatrix} 2 & 0 & 5 \\ 6 & 1 & 4 \\ -2 & 1 & 1 \end{bmatrix}\begin{bmatrix} 1 & 0 \\ -0 & 1 \\ 2 & 3 \end{bmatrix} =$

$$\begin{bmatrix} 2(1)+0(-1)+5(2) & 2(0)+0(1)+5(3) \\ 6(1)+1(-1)+4(2) & 6(0)+1(1)+4(3) \\ -2(1)+1(-1)+(2) & -2(0)+1(1)+1(3) \end{bmatrix} = \begin{bmatrix} 12 & 15 \\ 13 & 13 \\ -1 & 4 \end{bmatrix}$$

$\quad BA$ not possible

21. $AX = \begin{bmatrix} 2 & 1 \\ 1 & -1 \end{bmatrix} \begin{bmatrix} 1 \\ -1 \end{bmatrix} = \begin{bmatrix} 2(1)+1(-1) \\ 1(1)-1(-1) \end{bmatrix} = \begin{bmatrix} 1 \\ 2 \end{bmatrix} = B$

23. $AX = \begin{bmatrix} 2 & 1 & 1 \\ 1 & -1 & 1 \\ 1 & -2 & 2 \end{bmatrix} \begin{bmatrix} 1 \\ 1 \\ 3 \end{bmatrix} = \begin{bmatrix} 2(1)+1(1)+1(3) \\ 1(1)-1(1)+1(3) \\ 1(1)-2(1)+2(3) \end{bmatrix} = \begin{bmatrix} 6 \\ 3 \\ 5 \end{bmatrix} = B$

25. $AX = \begin{bmatrix} 2 & 1 & 3 \\ 1 & -2 & -1 \end{bmatrix} \begin{bmatrix} -\frac{1}{2} \\ 1 \\ 1 \end{bmatrix} = \begin{bmatrix} 2(-\frac{1}{2})+1(1)+3(1) \\ 1(-\frac{1}{2})-2(1)-1(1) \end{bmatrix} = \begin{bmatrix} 3 \\ -\frac{7}{2} \end{bmatrix} = B$

27. $BC = \begin{bmatrix} 0 & 1 \\ 5 & 2 \end{bmatrix} \begin{bmatrix} 3 & 5 \\ 0 & 2 \end{bmatrix} = \begin{bmatrix} 0 & 2 \\ 15 & 29 \end{bmatrix}$

$A(BC) = \begin{bmatrix} 1 & 2 \\ -1 & 0 \end{bmatrix} \begin{bmatrix} 0 & 2 \\ 15 & 29 \end{bmatrix} = \begin{bmatrix} 30 & 60 \\ 0 & -2 \end{bmatrix}$

$AB = \begin{bmatrix} 1 & 2 \\ -1 & 0 \end{bmatrix} \begin{bmatrix} 0 & 1 \\ 5 & 2 \end{bmatrix} = \begin{bmatrix} 10 & 5 \\ 0 & -1 \end{bmatrix}$

$(AB)C = \begin{bmatrix} 10 & 5 \\ 0 & -1 \end{bmatrix} \begin{bmatrix} 3 & 5 \\ 0 & 2 \end{bmatrix} = \begin{bmatrix} 30 & 60 \\ 0 & -2 \end{bmatrix}$

29. $A^2 = \begin{bmatrix} 1 & 2 \\ -1 & 0 \end{bmatrix} \begin{bmatrix} 1 & 2 \\ -1 & 0 \end{bmatrix} = \begin{bmatrix} -1 & 2 \\ -1 & -2 \end{bmatrix}$

$B^2 = \begin{bmatrix} 0 & 1 \\ 5 & 2 \end{bmatrix} \begin{bmatrix} 0 & 1 \\ 5 & 2 \end{bmatrix} = \begin{bmatrix} 5 & 2 \\ 10 & 9 \end{bmatrix}$

$A^2 - B^2 = \begin{bmatrix} -1 & 2 \\ -1 & -2 \end{bmatrix} - \begin{bmatrix} 5 & 2 \\ 10 & 9 \end{bmatrix} = \begin{bmatrix} -6 & 0 \\ -11 & -11 \end{bmatrix}$

$A - B = \begin{bmatrix} 1 & 2 \\ -1 & 0 \end{bmatrix} - \begin{bmatrix} 0 & 1 \\ 5 & 2 \end{bmatrix} = \begin{bmatrix} 1 & 1 \\ -6 & -2 \end{bmatrix}$

$A + B = \begin{bmatrix} 1 & 2 \\ -1 & 0 \end{bmatrix} + \begin{bmatrix} 0 & 1 \\ 5 & 2 \end{bmatrix} = \begin{bmatrix} 1 & 3 \\ 4 & 2 \end{bmatrix}$

$(A-B)(A+B) = \begin{bmatrix} 1 & 1 \\ -6 & -2 \end{bmatrix} \begin{bmatrix} 1 & 3 \\ 4 & 2 \end{bmatrix} = \begin{bmatrix} 5 & 5 \\ -14 & -22 \end{bmatrix}$

31. $(A+B)^2 = (A+B)(A+B) = A(A+B) + B(A+B) = A^2 + AB + BA + B^2$

33. $AB = \begin{bmatrix} 0 & 1 \\ 0 & 1 \end{bmatrix} \begin{bmatrix} 1 & 1 \\ 1 & 0 \end{bmatrix} = \begin{bmatrix} 1 & 0 \\ 1 & 0 \end{bmatrix}$

$AC = \begin{bmatrix} 0 & 1 \\ 0 & 1 \end{bmatrix} \begin{bmatrix} 0 & 0 \\ 1 & 0 \end{bmatrix} = \begin{bmatrix} 1 & 0 \\ 1 & 0 \end{bmatrix}$

Section 10.5   Solving Systems of Equations by Matrix Inversion

1. $\left[\begin{array}{cc|cc} 0 & 1 & 1 & 0 \\ 1 & 0 & 0 & 1 \end{array} \right] \xrightarrow{①\leftrightarrow②} \left[\begin{array}{cc|cc} 1 & 0 & 0 & 1 \\ 0 & 1 & 1 & 0 \end{array} \right]$

3. $\begin{bmatrix} 2 & 1 & | & 1 & 0 \\ 1 & 0 & | & 0 & 1 \end{bmatrix} \xrightarrow{① \leftrightarrow ②} \begin{bmatrix} 1 & 0 & | & 0 & 1 \\ 2 & 1 & | & 1 & 0 \end{bmatrix} \xrightarrow{-2①+②}$

$\begin{bmatrix} 1 & 0 & | & 0 & 1 \\ 0 & 1 & | & 1 & -2 \end{bmatrix}$

5. $\begin{bmatrix} 2 & 0 & 0 & | & 1 & 0 & 0 \\ 0 & 3 & 0 & | & 0 & 1 & 0 \\ 0 & 0 & 5 & | & 0 & 0 & 1 \end{bmatrix} \xrightarrow[\substack{\frac{1}{3}② \\ \frac{1}{5}③}]{\frac{1}{2}①} \begin{bmatrix} 1 & 0 & 0 & | & \frac{1}{2} & 0 & 0 \\ 0 & 1 & 0 & | & 0 & \frac{1}{3} & 0 \\ 0 & 0 & 1 & | & 0 & 0 & \frac{1}{5} \end{bmatrix}$

7. $\begin{bmatrix} 1 & 0 & 1 & | & 1 & 0 & 0 \\ 2 & 1 & 2 & | & 0 & 1 & 0 \\ -1 & 0 & 0 & | & 0 & 0 & 1 \end{bmatrix} \xrightarrow[\substack{-2①+②}]{①+③} \begin{bmatrix} 1 & 0 & 1 & | & 1 & 0 & 0 \\ 0 & 1 & 0 & | & -2 & 1 & 0 \\ 0 & 0 & 1 & | & 1 & 0 & 1 \end{bmatrix} \xrightarrow{-③+①}$

$\begin{bmatrix} 1 & 0 & 0 & | & 0 & 0 & -1 \\ 0 & 1 & 0 & | & -2 & 1 & 0 \\ 0 & 0 & 1 & | & 1 & 0 & 1 \end{bmatrix}$

11. $\begin{bmatrix} a & b & | & 1 & 0 \\ 0 & d & | & 0 & 1 \end{bmatrix} \xrightarrow[\frac{1}{d}②]{\frac{1}{a}①} \begin{bmatrix} 1 & \frac{b}{a} & | & \frac{1}{a} & 0 \\ 0 & 1 & | & 0 & \frac{1}{d} \end{bmatrix} \xrightarrow{\frac{-b}{a}②+①} \begin{bmatrix} 1 & 0 & | & \frac{1}{a} & -\frac{b}{ad} \\ 0 & 1 & | & 0 & \frac{1}{d} \end{bmatrix}$

13.

$$\left[\begin{array}{ccc|ccc} 2 & 4 & 3 & 1 & 0 & 0 \\ -2 & 4 & -2 & 0 & 1 & 0 \\ 1 & 0 & 1 & 0 & 0 & 1 \end{array}\right] \xrightarrow{\text{①}+\text{②}} \left[\begin{array}{ccc|ccc} 2 & 4 & 3 & 1 & 0 & 0 \\ 0 & 8 & 1 & 1 & 1 & 0 \\ 1 & 0 & 1 & 0 & 0 & 1 \end{array}\right] \xrightarrow{-\text{③}+\text{①}}$$

$$\left[\begin{array}{ccc|ccc} 1 & 4 & 2 & 1 & 0 & -1 \\ 0 & 8 & 1 & 1 & 1 & 0 \\ 1 & 0 & 1 & 0 & 0 & 1 \end{array}\right] \xrightarrow{-\text{①}+\text{③}} \left[\begin{array}{ccc|ccc} 1 & 4 & 2 & 1 & 0 & -1 \\ 0 & 8 & 1 & 1 & 1 & 0 \\ 0 & -4 & -1 & -1 & 0 & 2 \end{array}\right] \xrightarrow[-\frac{1}{4}\text{③}]{\frac{1}{8}\text{②}}$$

$$\left[\begin{array}{ccc|ccc} 1 & 4 & 2 & 1 & 0 & -1 \\ 0 & 1 & \frac{1}{8} & \frac{1}{8} & \frac{1}{8} & 0 \\ 0 & 1 & \frac{1}{4} & \frac{1}{4} & 0 & -\frac{1}{2} \end{array}\right] \xrightarrow{-\text{②}+\text{③}} \left[\begin{array}{ccc|ccc} 1 & 4 & 2 & 1 & 0 & -1 \\ 0 & 1 & \frac{1}{8} & \frac{1}{8} & \frac{1}{8} & 0 \\ 0 & 0 & \frac{1}{8} & \frac{1}{8} & -\frac{1}{8} & -\frac{1}{2} \end{array}\right] \xrightarrow[-4\text{②}+\text{①}]{8\text{③}}$$

$$\left[\begin{array}{ccc|ccc} 1 & 0 & \frac{3}{2} & \frac{1}{2} & -\frac{1}{2} & -1 \\ 0 & 1 & \frac{1}{8} & \frac{1}{8} & \frac{1}{8} & 0 \\ 0 & 0 & 1 & 1 & -1 & -4 \end{array}\right] \xrightarrow[-\frac{1}{8}\text{③}+\text{②}]{-\frac{3}{2}\text{③}+\text{①}} \left[\begin{array}{ccc|ccc} 1 & 0 & 0 & -1 & 1 & 5 \\ 0 & 1 & 0 & 0 & \frac{1}{4} & \frac{1}{2} \\ 0 & 0 & 1 & 1 & -1 & -4 \end{array}\right]$$

15.

$$\left[\begin{array}{ccc|ccc} -1 & 2 & 2 & 1 & 0 & 0 \\ -1 & 2 & 5 & 0 & 1 & 0 \\ -2 & 2 & 10 & 0 & 0 & 1 \end{array}\right] \xrightarrow[-2\text{①}+\text{③}]{-\text{①}+\text{②}} \left[\begin{array}{ccc|ccc} -1 & 2 & 2 & 1 & 0 & 0 \\ 0 & 0 & 3 & -1 & 1 & 0 \\ 0 & -2 & 6 & -2 & 0 & 1 \end{array}\right] \xrightarrow[\text{②}\leftrightarrow\text{③}]{-\text{①}}$$

$$\left[\begin{array}{ccc|ccc} 1 & -2 & -2 & -1 & 0 & 0 \\ 0 & -2 & 6 & -2 & 0 & 1 \\ 0 & 0 & 3 & -1 & 1 & 0 \end{array}\right] \xrightarrow[\frac{1}{3}\text{③}]{-\frac{1}{2}\text{②}} \left[\begin{array}{ccc|ccc} 1 & -2 & -2 & -1 & 0 & 0 \\ 0 & 1 & -3 & 1 & 0 & -\frac{1}{2} \\ 0 & 0 & 1 & -\frac{1}{3} & \frac{1}{3} & 0 \end{array}\right] \xrightarrow[2\text{③}+\text{①}]{3\text{③}+\text{②}}$$

$$\left[\begin{array}{ccc|ccc} 1 & -2 & 0 & -\frac{5}{3} & \frac{2}{3} & 0 \\ 0 & 1 & 0 & 0 & 1 & -\frac{1}{2} \\ 0 & 0 & 1 & -\frac{1}{3} & \frac{1}{3} & 0 \end{array}\right] \xrightarrow{2\text{②}+\text{①}} \left[\begin{array}{ccc|ccc} 1 & 0 & 0 & -\frac{5}{3} & \frac{8}{3} & -1 \\ 0 & 1 & 0 & 0 & 1 & -\frac{1}{2} \\ 0 & 0 & 1 & -\frac{1}{3} & \frac{1}{3} & 0 \end{array}\right]$$

17. $\begin{bmatrix} 2 & 1 \\ 1 & -1 \end{bmatrix} \begin{bmatrix} x \\ y \end{bmatrix} = \begin{bmatrix} 4 \\ 0 \end{bmatrix}$

$\left[\begin{array}{cc|cc} 2 & 1 & 1 & 0 \\ 1 & -1 & 0 & 1 \end{array}\right] \xrightarrow{-②+①} \left[\begin{array}{cc|cc} 1 & 2 & 1 & 1 \\ 1 & -1 & 0 & 1 \end{array}\right] \xrightarrow{-①+②}$

$\left[\begin{array}{cc|cc} 1 & 2 & 1 & 1 \\ 0 & -3 & -1 & 0 \end{array}\right] \xrightarrow{-\frac{1}{3}②} \left[\begin{array}{cc|cc} 1 & 2 & 1 & 1 \\ 0 & 1 & \frac{1}{3} & 0 \end{array}\right] \xrightarrow{-2②+①}$

$\left[\begin{array}{cc|cc} 1 & 0 & \frac{1}{3} & 1 \\ 0 & 1 & \frac{1}{3} & 0 \end{array}\right]$

$\begin{bmatrix} \frac{1}{3} & 1 \\ \frac{1}{3} & 0 \end{bmatrix} \begin{bmatrix} 4 \\ 0 \end{bmatrix} = \begin{bmatrix} \frac{4}{3} \\ \frac{4}{3} \end{bmatrix}$

19. $\begin{bmatrix} 2 & 0 & 1 \\ 0 & 3 & 1 \\ -2 & 0 & 4 \end{bmatrix} \begin{bmatrix} x \\ y \\ z \end{bmatrix} = \begin{bmatrix} 1 \\ 2 \\ 0 \end{bmatrix}$

$\left[\begin{array}{ccc|ccc} 2 & 0 & 1 & 1 & 0 & 0 \\ 0 & 3 & 1 & 0 & 1 & 0 \\ -2 & 0 & 4 & 0 & 0 & 1 \end{array}\right] \xrightarrow[\frac{1}{3}②]{①+③} \left[\begin{array}{ccc|ccc} 2 & 0 & 1 & 1 & 0 & 0 \\ 0 & 1 & \frac{1}{3} & 0 & \frac{1}{3} & 0 \\ 0 & 0 & 5 & 1 & 0 & 1 \end{array}\right] \xrightarrow[\frac{1}{5}③]{\frac{1}{2}①}$

$\left[\begin{array}{ccc|ccc} 1 & 0 & \frac{1}{2} & \frac{1}{2} & 0 & 0 \\ 0 & 1 & \frac{1}{3} & 0 & \frac{1}{3} & 0 \\ 0 & 0 & 1 & \frac{1}{5} & 0 & \frac{1}{5} \end{array}\right] \xrightarrow[-\frac{1}{3}③+②]{-\frac{1}{2}③+①} \left[\begin{array}{ccc|ccc} 1 & 0 & 0 & \frac{2}{5} & 0 & -\frac{1}{10} \\ 0 & 1 & 0 & -\frac{1}{15} & \frac{1}{3} & -\frac{1}{15} \\ 0 & 0 & 1 & \frac{1}{5} & 0 & \frac{1}{5} \end{array}\right]$

$\begin{bmatrix} \frac{2}{5} & 0 & -\frac{1}{10} \\ -\frac{1}{15} & \frac{1}{3} & -\frac{1}{15} \\ \frac{1}{5} & 0 & \frac{1}{5} \end{bmatrix} \begin{bmatrix} 1 \\ 2 \\ 0 \end{bmatrix} = \begin{bmatrix} \frac{2}{5} \\ \frac{3}{5} \\ \frac{1}{5} \end{bmatrix}$

21.
$$\begin{bmatrix} 1 & 0 & 1 \\ 2 & 1 & 2 \\ -1 & 0 & 0 \end{bmatrix} \begin{bmatrix} x \\ y \\ z \end{bmatrix} = \begin{bmatrix} - \\ 2 \\ 1 \end{bmatrix}$$

$$A^{-1} = \begin{bmatrix} 0 & 0 & -1 \\ -2 & 1 & 0 \\ 1 & 0 & 1 \end{bmatrix} \qquad \text{(See 10.5, \#7)}$$

$$\begin{bmatrix} 0 & 0 & -1 \\ -2 & 1 & 0 \\ 1 & 0 & 1 \end{bmatrix} \begin{bmatrix} 0 \\ 2 \\ 1 \end{bmatrix} = \begin{bmatrix} -1 \\ 2 \\ 1 \end{bmatrix}$$

23.
$$\begin{bmatrix} -1 & 2 & 2 \\ -1 & 2 & 5 \\ -1 & 1 & 5 \end{bmatrix} \begin{bmatrix} x \\ y \\ z \end{bmatrix} = \begin{bmatrix} 0 \\ 0 \\ -2 \end{bmatrix}$$

Multiply row 3 by 2,

$$\begin{bmatrix} -1 & 2 & 2 \\ -1 & 2 & 5 \\ -2 & 2 & 10 \end{bmatrix} \begin{bmatrix} x \\ y \\ z \end{bmatrix} = \begin{bmatrix} 0 \\ 0 \\ -4 \end{bmatrix}$$

$$A^{-1} = \begin{bmatrix} -\frac{5}{3} & \frac{8}{3} & -1 \\ 0 & 1 & -\frac{1}{2} \\ -\frac{1}{3} & \frac{1}{3} & 0 \end{bmatrix} \qquad \text{(See 10.5, \#15)}$$

$$\begin{bmatrix} -\frac{5}{3} & \frac{8}{3} & -1 \\ 0 & 1 & -\frac{1}{2} \\ -\frac{1}{3} & \frac{1}{3} & 0 \end{bmatrix} \begin{bmatrix} 0 \\ 0 \\ -4 \end{bmatrix} = \begin{bmatrix} 4 \\ 2 \\ 0 \end{bmatrix}$$

25.
$$\begin{bmatrix} 1 & 1 & 5 \\ 3 & 8 & 5 \\ -1 & -5 & 3 \end{bmatrix} \begin{bmatrix} x \\ y \\ z \end{bmatrix} = \begin{bmatrix} 2 \\ 0 \\ 15 \end{bmatrix}$$

$$\left[\begin{array}{ccc|ccc} 1 & 1 & 5 & 1 & 0 & 0 \\ 3 & 8 & 5 & 0 & 1 & 0 \\ -1 & -5 & 3 & 0 & 0 & 1 \end{array}\right] \xrightarrow[\text{①}+\text{③}]{-3\text{①}+\text{②}} \left[\begin{array}{ccc|ccc} 1 & 1 & 5 & 1 & 0 & 0 \\ 0 & 5 & 10 & -3 & 1 & 0 \\ 0 & -4 & 8 & 1 & 0 & 1 \end{array}\right] \xrightarrow{\frac{1}{5}\text{②}}$$

$$\left[\begin{array}{ccc|ccc} 1 & 1 & 5 & 1 & 0 & 0 \\ 0 & 1 & 2 & -\frac{3}{5} & \frac{1}{5} & 0 \\ 0 & -4 & 8 & 1 & 0 & 1 \end{array}\right] \xrightarrow{4\text{②}+\text{③}} \left[\begin{array}{ccc|ccc} 1 & 1 & 5 & 1 & 0 & 0 \\ 0 & 1 & -2 & -\frac{3}{5} & \frac{1}{5} & 0 \\ 0 & 0 & 0 & -\frac{7}{5} & -\frac{4}{5} & 1 \end{array}\right]$$

No solution

Section 10.6   Determinants

1. $\begin{vmatrix} 2 & 4 \\ 3 & 5 \end{vmatrix} = 10 - 12 = -2$

3. $\begin{vmatrix} 0 & 1 \\ 2 & 6 \end{vmatrix} = 0 - 2 = -2$

5. $\begin{vmatrix} 2 & -6 \\ 5 & -3 \end{vmatrix} = -6 + 30 = 24$

7. $\begin{vmatrix} 5 & 10 \\ 2 & 4 \end{vmatrix} = 20 - 20 = 0$

9. $\begin{vmatrix} 2 & 0 & 0 \\ 3 & 2 & 4 \\ 1 & -3 & 5 \end{vmatrix} = 20 + 0 + 0 - (0-24+0) = 44$

$2\begin{vmatrix} 2 & 4 \\ -3 & 5 \end{vmatrix} = 2(10+12) = 44$

11.
$$\begin{vmatrix} 2 & 3 & 4 \\ 2 & 0 & 3 \\ -1 & 6 & 5 \end{vmatrix} = 0 - 9 - 48 - (0+36+30) = -123$$

By row 2:

$$-2\begin{vmatrix} 3 & -4 \\ 6 & 5 \end{vmatrix} \quad -3\begin{vmatrix} 2 & 3 \\ -1 & 6 \end{vmatrix} = -2(15+24) - 3(12+3) = -123$$

13.
$$\begin{vmatrix} 3 & 1 & 9 \\ 6 & 1 & -2 \\ -4 & 1 & 5 \end{vmatrix} = 15 + 8 + 54 - (-36-6+30) = 89$$

By column 2:

$$-1\begin{vmatrix} 6 & -2 \\ -4 & 5 \end{vmatrix} \quad +1\begin{vmatrix} 3 & 9 \\ -4 & 5 \end{vmatrix} \quad -1\begin{vmatrix} 3 & 9 \\ 6 & -2 \end{vmatrix}$$

$$= -(30-8) + (15+36) - (-6-54) = 89$$

15.
$$\begin{vmatrix} 6 & 7 & 6 \\ 2 & 5 & -2 \\ 3 & 1 & -3 \end{vmatrix} = -90 + 42 + 12 - (90+12-42) = -96$$

By row 3:

$$3\begin{vmatrix} 7 & 6 \\ 5 & 2 \end{vmatrix} \quad -1\begin{vmatrix} 6 & 6 \\ 2 & 2 \end{vmatrix} \quad -3\begin{vmatrix} 6 & 7 \\ 2 & 5 \end{vmatrix}$$

$$= 3(14-30) -1(12-12) -3(30-14) = -96$$

17.
$$\begin{vmatrix} 9 & -6 & 0 \\ -1 & -3 & -1 \\ -5 & -4 & -7 \end{vmatrix} = 189 - 30 + 0 - (0+36-42) = 165$$

By row 1:

$$9\begin{vmatrix} -3 & -1 \\ -4 & -7 \end{vmatrix} \quad +6\begin{vmatrix} -1 & -1 \\ -5 & -7 \end{vmatrix} = 9(21-4) +6(7-5) = 165$$

19.

$$\begin{vmatrix} x & 1 \\ 1 & x \end{vmatrix} = 0$$

$$x^2 - 1 = 0$$

$$x = \pm 1$$

21.

$$\begin{Vmatrix} \begin{vmatrix} 3 & 1 \\ -1 & 2 \\ 5 & 0 \\ 0 & 1 \end{vmatrix} & \begin{vmatrix} 7 & 0 \\ 1 & 2 \\ 0 & 0 \\ 0 & 1 \end{vmatrix} \end{Vmatrix} = \begin{vmatrix} 7 & 14 \\ 5 & 0 \end{vmatrix} = -70$$

23.

$$\begin{vmatrix} x & 2 & 1 \\ -1 & 3 & 0 \\ 1 & x & 0 \end{vmatrix} = 0$$

By column 3:

$$0 = 1 \begin{vmatrix} -1 & 3 \\ 1 & x \end{vmatrix} = -x - 3$$

$$x = -3$$

25.

$$\begin{vmatrix} 1 & 4 & -2 & 0 \\ 0 & 1 & 1 & 2 \\ -3 & -2 & 1 & 0 \\ -1 & 1 & 2 & 3 \end{vmatrix} = 2 \begin{vmatrix} 1 & 4 & -2 \\ -3 & -2 & 1 \\ -1 & 1 & 2 \end{vmatrix} + 3 \begin{vmatrix} 1 & 4 & -2 \\ 0 & 1 & 1 \\ -3 & -2 & 1 \end{vmatrix}$$

$$= \left[-4 - 4 + 6 - (-4 + 1 - 24) \right] + 3 \left[1 + 0 - 12 - (6 - 2 + 0) \right] = 5$$

27a. If a row (col) of A has all zeros, then $\det A = 0$.

$$\det A = \begin{vmatrix} a_{11} & 0 \\ a_{21} & 0 \end{vmatrix} = a_{11}(0) - a_{21}(0) = 0$$

27b. If A is obtained by interchanging two rows (cols), then $\det A = -\det B$.

$$\det B = \begin{vmatrix} a_{21} & a_{22} \\ a_{11} & a_{12} \end{vmatrix} = a_{21}a_{12} - a_{11}a_{22} = -(a_{11}a_{22} - a_{21}a_{12}) = - \begin{vmatrix} a_{11} & a_{12} \\ a_{21} & a_{22} \end{vmatrix}$$

$$= -\det A$$

27c. If any two rows (cols) of A are equal, then $\det A = 0$.

$$\begin{vmatrix} a_{11} & a_{12} \\ a_{11} & a_{12} \end{vmatrix} = a_{11}a_{12} - a_{11}a_{12} = 0$$

27d. If A and B are identical except that a row (col) of B is k times the same row (col) of A, then $\det A = k \det B$.

$$\det B = \begin{vmatrix} ka_{11} & ka_{12} \\ a_{21} & a_{22} \end{vmatrix} = ka_{11}a_{22} - ka_{21}a_{12} = k(a_{11}a_{22} - a_{21}a_{12}) = k \det A$$

27e. The value of the determinant is unchanged if the rows and columns of the matrix are interchanged.

$$\begin{vmatrix} a_{11} & a_{12} \\ a_{21} & a_{22} \end{vmatrix} = a_{11}a_{22} - a_{12}a_{21} = \begin{vmatrix} a_{11} & a_{21} \\ a_{12} & a_{22} \end{vmatrix}$$

27f. The value of a determinant is unchanged if a multiple of one row (col) is added to another row (col.)

$$\det B = \begin{vmatrix} a_{11} & a_{12} \\ ka_{11}+a_{21} & ka_{12}+a_{22} \end{vmatrix} = a_{11}(ka_{12}+a_{22}) - a_{12}(ka_{11}+a_{21})$$

$$= a_{11}a_{22} - a_{12}a_{21} = \det A$$

1. $x = \dfrac{\begin{vmatrix} 2 & -4 \\ -4 & 1 \end{vmatrix}}{\begin{vmatrix} 1 & -4 \\ -2 & 1 \end{vmatrix}} = \dfrac{-14}{-9} = \dfrac{14}{9}$ $y = \dfrac{\begin{vmatrix} 1 & 2 \\ -2 & -4 \end{vmatrix}}{-9} = 0$

3. $x = \dfrac{\begin{vmatrix} 23 & 4 \\ -1 & -3 \end{vmatrix}}{\begin{vmatrix} 3 & 4 \\ 1 & -3 \end{vmatrix}} = \dfrac{-65}{-13} = 5,$ $y = \dfrac{\begin{vmatrix} 3 & 23 \\ 1 & -1 \end{vmatrix}}{-13} = \dfrac{-26}{-13} = 2$

5. $x = \dfrac{\begin{vmatrix} 8 & -4 \\ 16 & -8 \end{vmatrix}}{\begin{vmatrix} 1 & -4 \\ 2 & -8 \end{vmatrix}} = \dfrac{0}{0},$ Dependent

7. $x = \dfrac{\begin{vmatrix} 10 & 4 \\ -2 & 8 \end{vmatrix}}{\begin{vmatrix} 3 & 4 \\ 6 & 8 \end{vmatrix}} = \dfrac{88}{0},$ Inconsistent

9. $x = \dfrac{\begin{vmatrix} 2 & -1 \\ 6 & 1 \end{vmatrix}}{\begin{vmatrix} 3 & -1 \\ 2 & 1 \end{vmatrix}} = \dfrac{8}{5},$ $y = \dfrac{\begin{vmatrix} 3 & 2 \\ 2 & 6 \end{vmatrix}}{5} = \dfrac{14}{5}$

11. $x = \dfrac{\begin{vmatrix} 2 & -1 \\ -5 & 1 \end{vmatrix}}{\begin{vmatrix} 2 & -1 \\ 1 & 1 \end{vmatrix}} = \dfrac{-3}{3} = -1, \qquad y = \dfrac{\begin{vmatrix} 2 & 2 \\ 1 & -5 \end{vmatrix}}{3} = \dfrac{-12}{3} = -4$

13. $x = \dfrac{\begin{vmatrix} 8 & 1 & 1 \\ 12 & 1 & -1 \\ 2 & -1 & 1 \end{vmatrix}}{\begin{vmatrix} 1 & 1 & 1 \\ 1 & 1 & -1 \\ 1 & -1 & 1 \end{vmatrix}} = \dfrac{-28}{-4} = 7, \qquad y = \dfrac{\begin{vmatrix} 1 & 8 & 1 \\ 1 & 12 & -1 \\ 1 & 2 & 1 \end{vmatrix}}{-4} = \dfrac{-12}{-4} = 3,$

$$z = \dfrac{\begin{vmatrix} 1 & 1 & 8 \\ 1 & 1 & 12 \\ 1 & -1 & 2 \end{vmatrix}}{-4} = \dfrac{8}{-4} = -2$$

15. $r = \dfrac{\begin{vmatrix} 2 & 3 & 2 \\ -1 & 2 & -1 \\ 0 & -3 & 3 \end{vmatrix}}{\begin{vmatrix} 1 & 3 & 2 \\ 3 & 2 & -1 \\ 1 & -3 & -3 \end{vmatrix}} = \dfrac{21}{-7} = -3, \qquad s = \dfrac{\begin{vmatrix} 1 & 2 & 2 \\ 3 & -1 & -1 \\ 1 & 0 & 3 \end{vmatrix}}{-7} = \dfrac{-21}{-7} = 3$

$$t = \dfrac{\begin{vmatrix} 1 & 3 & 2 \\ 3 & 2 & -1 \\ 1 & -3 & 0 \end{vmatrix}}{-7} = \dfrac{-28}{-7} = 4$$

17. $x = \dfrac{\begin{vmatrix} 5 & -3 & 4 \\ 4 & -7 & 4 \\ 2 & 3 & -5 \end{vmatrix}}{\begin{vmatrix} 2 & -3 & 4 \\ 5 & -7 & 4 \\ -2 & 3 & -5 \end{vmatrix}} = \dfrac{135}{-1} = -135, \qquad y = \dfrac{\begin{vmatrix} 2 & 5 & 4 \\ 5 & 4 & 4 \\ -2 & 2 & -5 \end{vmatrix}}{-1} = \dfrac{101}{1} = -101$

$$z = \dfrac{\begin{vmatrix} 2 & -3 & 5 \\ 5 & -7 & 4 \\ -2 & 3 & 2 \end{vmatrix}}{-1} = \dfrac{7}{-1} = -7$$

19. $x = \dfrac{\begin{vmatrix} 0 & 2 & -1 \\ 0 & -1 & 1 \\ 0 & -3 & -2 \end{vmatrix}}{\begin{vmatrix} 1 & 2 & -1 \\ 1 & -1 & 1 \\ 2 & -3 & -2 \end{vmatrix}} = \dfrac{0}{14} = 0, \quad y = 0, \quad z = 0$

21. $u = \dfrac{\begin{vmatrix} 8 & -4 & 2 \\ 1 & 2 & 1 \\ 17 & 0 & -18 \end{vmatrix}}{\begin{vmatrix} 9 & -4 & 2 \\ 3 & 2 & 1 \\ 12 & 0 & -18 \end{vmatrix}} = \dfrac{-496}{-636}, \qquad v = \dfrac{\begin{vmatrix} 9 & 8 & 2 \\ 3 & 1 & 1 \\ 12 & 17 & -18 \end{vmatrix}}{-636} = \dfrac{291}{-636}$

$$w = \dfrac{\begin{vmatrix} 9 & -4 & 8 \\ 3 & 2 & 1 \\ 12 & 0 & 17 \end{vmatrix}}{-636} = \dfrac{270}{-636}$$

23. $w + 6 = l,$ $\quad w - l = 6$

$2w + 2l = 40$

$$w = \dfrac{\begin{vmatrix} 6 & -1 \\ 40 & 2 \end{vmatrix}}{\begin{vmatrix} 1 & -1 \\ 2 & 2 \end{vmatrix}} = \frac{52}{4} = 13 \qquad l = \dfrac{\begin{vmatrix} 1 & 6 \\ 2 & 40 \end{vmatrix}}{4} = 7$$

25.

	Rate	Time
R_1	50	$t+1$
R_2	60	t

$50(t+1) = d,$ $\quad 50t - d = -50$

$60t \quad = d,$ $\quad 60t - d = \quad 0$

$$t = \dfrac{\begin{vmatrix} -50 & -1 \\ 0 & -1 \end{vmatrix}}{\begin{vmatrix} 50 & -1 \\ 60 & -1 \end{vmatrix}} = \frac{50}{10} = 5$$

27. $5n + 10d + 25q = 1900$

$\quad n + \quad d + \quad q = 100$

$\quad n - \quad 3d \qquad = \quad 0$

$$d = \dfrac{\begin{vmatrix} 5 & 1900 & 25 \\ 1 & 100 & 1 \\ 1 & 0 & 0 \end{vmatrix}}{\begin{vmatrix} 5 & 10 & 25 \\ 1 & 1 & 1 \\ 1 & -3 & 0 \end{vmatrix}} = \frac{-600}{-65} = 8$$

29.
$$x + y + z = 50 \qquad x + y + z = 50$$
$$x = y + 5 \qquad\quad x - y = 5$$
$$x = z + 10 \qquad\quad x - z = 10$$

$$x = \dfrac{\begin{vmatrix} 50 & 1 & 1 \\ 5 & -1 & 0 \\ 10 & 0 & -1 \end{vmatrix}}{\begin{vmatrix} 1 & 1 & 1 \\ 1 & -1 & 0 \\ 1 & 0 & -1 \end{vmatrix}} = \dfrac{65}{3}, \qquad y = \dfrac{\begin{vmatrix} 1 & 50 & 1 \\ 1 & 5 & 0 \\ 1 & 10 & -1 \end{vmatrix}}{3} = \dfrac{50}{3}$$

$$z = \dfrac{\begin{vmatrix} 1 & 1 & 50 \\ 1 & -1 & 5 \\ 1 & 0 & 10 \end{vmatrix}}{3} = \dfrac{35}{3}$$

31.
$$u' = \dfrac{\begin{vmatrix} F & Q_1 \\ 0 & Q_2 \end{vmatrix}}{\begin{vmatrix} R_1 & Q_1 \\ R_2 & Q_2 \end{vmatrix}} = \dfrac{FQ_2}{R_1 Q_2 - R_2 Q_1}$$

$$v' = \dfrac{\begin{vmatrix} R_1 & F \\ R_2 & 0 \end{vmatrix}}{\begin{vmatrix} R_1 & Q_1 \\ R_2 & Q_2 \end{vmatrix}} = \dfrac{-R_2 F}{R_1 Q_2 - R_2 Q_1}$$

33. $(1, 0) \to 0 = 1 + a + b + c$ $\qquad$ $a + b = -2$

$\quad$ $(3, 0) \to 0 = 27 + 9a + 3b + c$ $\qquad$ $9a + 3b = -28$

$\quad$ $(0, 1) \to 1 = c$

$$a = \dfrac{\begin{vmatrix} -2 & 1 \\ -28 & 3 \end{vmatrix}}{\begin{vmatrix} 1 & 1 \\ 9 & 3 \end{vmatrix}} = \dfrac{-22}{-6} = \dfrac{11}{3}, \qquad b = \dfrac{\begin{vmatrix} 1 & -2 \\ 9 & -28 \end{vmatrix}}{-6} = \dfrac{-10}{-6} = \dfrac{5}{3}$$

35. $(1, 2) \to 2 = \dfrac{a}{1+c}, \qquad (2, 1) \to 1 = \dfrac{2a}{2+c}$

$\quad$ $a - 2c = 2$

$\quad$ $2a - c = 2$

$$a = \dfrac{\begin{vmatrix} 2 & -2 \\ 2 & -1 \end{vmatrix}}{\begin{vmatrix} 1 & -2 \\ 2 & -1 \end{vmatrix}} = \dfrac{2}{3}, \qquad c = \dfrac{\begin{vmatrix} 1 & 2 \\ 2 & 2 \end{vmatrix}}{3} = \dfrac{-2}{3}$$

Section 10.8 $\quad$ Linear Inequalities in Two Unknowns

5. $\;\; x + y > x - y$

$\qquad 2y > 0$

$\qquad y > 0$

7. $2x - y - 8 > x - 2y + 8$

$\qquad x + y > 16$

9. $\;\; 3(x-1) + 4x \geq 7x + y + 2(x-y)$

$\qquad 3x - 3 + 4x \geq 7x + y + 2x - 2y$

11. $|y| < 3$

$\qquad y < 3 \qquad$ or $\qquad y > -3$

13. $|x| < y$

$\qquad x < y \qquad$ or $\qquad x > -y$

15. $|x| < 1-|y|$

    a. $x \geq 0$ and $y \geq 0$ **b.** $x \geq 0$ and $y < 0$

           $x < 1-y$ $x < 1+y$

    c. $x < 0$ and $y \geq 0$ **d.** $x < 0$ and $y < 0$

           $-x < 1-y$ $-x < 1+y$

           $x > y-1$ $x > -1 - y$

17. $\frac{9}{5}C + 32 < 0$ **19.** $35x + 15y \leq 1000$

        $\frac{9}{5}C < -32$ $x = 0, \quad y \leq \frac{1000}{15} \simeq 67$

        $C < -\frac{160}{9}$ $y = 0, \quad x \leq \frac{1000}{35} \simeq 29$

Sections 10.9 and 10.10   Systems of Linear Inequalities

                            Linear Programming

17. $y = 2$

     $x + 2y = 8$

     $x = 8-4 = 4, (4.2)$

     $p = x+3y$

     $(0, 2) \quad p = 0 + 3(2) = 6$

     $(0, 0) \quad p = 0 + 0 = 0$

     $(4, 2) \quad p = 4 + 3(2) = 10$, maximum

     $(8, 0) \quad p = 8 + 0 = 8$

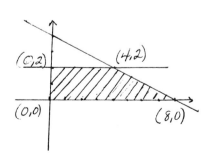

19. $x + y = 5$
$\underline{x + 2y = 8}$
$\quad - \quad y = -3, \quad y = 3, \quad x = 2$

$p = 5x + y$

$(0, 4) \quad p = 5(0) + 4 = 4$

$(0, 0) \quad p = 5(0) + 0 = 0$

$(2, 3) \quad p = 5(2) + 3 = 13$

$(5, 0) \quad p = 5(5) + 0 = 25$, maximum

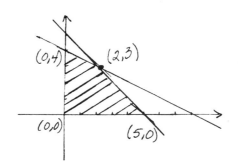

21. $\quad x + 3y = 8$
$\underline{2x + y = 6}$

$\quad 2x + y = \quad 6$
$\underline{-2x - 6y = -16}$
$\quad - \quad 5y = -10,$
$\quad\quad y = 2, \quad x = 2$

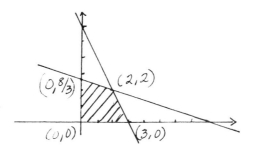

a. $p = 2x + 3y$

$\quad (0, 0) \quad p = 0 + 0 = 0$

$\quad (0, \frac{8}{3}) \quad p = 0 + 3(\frac{8}{3}) = 8$

$\quad (3, 0) \quad p = 2(3) + 0 = 6$

$\quad (2, 2) \quad p = 2(2) + 3(2) = 10$, maximum

b. $p = 3x + y$

$\quad (0, 0) \quad p = 0 + 0 = 0$

$\quad (0, \frac{8}{3}) \quad p = 3(0) + (\frac{8}{3}) = \frac{8}{3}$

$\quad (3, 0) \quad p = 3(3) + 9 = 9$, maximum

$\quad (2, 2) \quad p = 3(2) + 2 = 8$

23. $x + 2y = 3$
$\underline{x + 3y = 4}$
$\quad\quad\quad y = 1, \quad x = 1$

$p = 5x + 3y$

$(1, 1) \quad p = 5(1) + 3(1) = 8$

$(0, \frac{3}{2}) \quad p = 5(0) + 3(\frac{3}{2}) = \frac{9}{2}$, minimum

$(2, \frac{2}{3}) \quad p = 5(2) + 3(\frac{3}{2}) = 12$

25. $2x + y = 5$
$\underline{x + 2y = 4}$

$2x + y = 5$
$\underline{-2x - 4y = -8}$
$y = 1, \quad x = 2$

$p = 3x + y$

$(0, 2) \quad p = 3(0) + 2 = 2$

$(2, 1) \quad p = 3(2) + 1 = 7$

$(0, 0) \quad p = 0$

$(\frac{5}{2}, 0) \quad p = 3(\frac{5}{2}) + 0 = \frac{15}{2}$, maximum

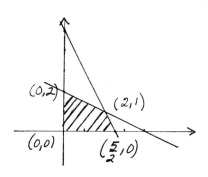

27. $2x + 3y = 120$
$\underline{4x + y = 90}$

$2x + 3y = 120$
$\underline{-12x - 3y = -270}$
$-10x = -150$
$x = 15, \quad y = 30$

$p = 10x + 8y$

$(0, 40) \quad p = 10(0) + 8(40) = 320$

$(15, 30) \quad p = 10(15) + 8(30) = 390$

$(\frac{45}{2}, 0) \quad p = 10(\frac{45}{2}) + 0 = 225$

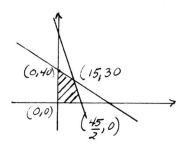

29. $x \geq 0, \quad y \geq 0, \quad x + y \leq 20{,}000$

$y \leq 2x,$

$I = .07x + .085y$

$x + y = 20{,}000$
$\underline{2x - y = 0}$
$3x = 20{,}000, \qquad x = 6700$
$\phantom{x + 3x = 20{,}000,}y = 13{,}300$

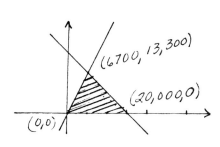

$(0, 0) \qquad\qquad I = 0$

$(20{,}000, 0) \qquad I = .07(20{,}000) + 0 = 1400$

$(6700, 13{,}300) \qquad I = .07(6700) + .085(13{,}300) = 1179.5$

1. $x + 3y = 5$
    $\underline{2x - 3y = 1}$

    $3x = 6, \quad x = 2$
    $3y = 5 - x = 3, \quad y = 1$

3. $2s + 5t = -4$
        $\underline{6s + 8t = 10}$

        $-6s - 15t = 12$
        $\underline{6s + 8t = 10}$

        $-7t = 22, \quad t = -\frac{22}{7}$

        $2s = -4 - 5\left(-\frac{22}{7}\right), \quad s = \frac{41}{7}$

5. $5x + 6z + 3 = 0$
        $\underline{7x + 4z - 9 = 0}$

        $10x + 12z + 6 = 0$
        $\underline{-21x - 12z + 27 = 0}$

        $-11x = -33, \quad x = 3$
        $6z = -5x - 3, \quad z = -3$

7. $x + y + z = 1$
    $x + y - z = 2$
    $\underline{x - y + z = 3}$

    $x + y + z = 1$
    $\underline{x + y - z = 2}$
    $2x + 2y = 3$

    $x + y - z = 2$
    $\underline{x - y + z = 3}$
    $2x = 5, \quad x = \frac{5}{2}$

    $2y = 3 - 2x = 3 - 2\left(\frac{5}{2}\right),$
    $y = -1$
    $z = 1 - 1 - y = 1 - \frac{5}{2} + 1$
    $ = -\frac{1}{2}$

9. $a + 2b - 3c = 1$
        $b - c = 4$
        $\underline{-a + b + c = 5}$

        $a + 2b - 3c = 1$
        $\underline{-a + b + c = 5}$
        $3b - 2c = 6$
        $\underline{2b - 2c = 8}$

        $b = -2$
        $c = b - 4 = -2 - 4 = -6$
        $a = b + c - 5 = -2 - 6 - 5 = -13$

11. $\begin{bmatrix} 3 & 5 & \vdots & 13 \\ -1 & 3 & \vdots & 3 \end{bmatrix} \xrightarrow{\frac{1}{3}①+②} \begin{bmatrix} 3 & 5 & \vdots & 13 \\ 0 & \frac{14}{3} & \vdots & \frac{28}{3} \end{bmatrix} \xrightarrow{\frac{3}{14}②} \begin{bmatrix} 3 & 5 & \vdots & 13 \\ 0 & 1 & \vdots & 2 \end{bmatrix} \xrightarrow{-5②+①}$

$\begin{bmatrix} 3 & 0 & \vdots & 3 \\ 0 & 1 & \vdots & 2 \end{bmatrix} \xrightarrow{\frac{1}{3}①} \begin{bmatrix} 1 & 0 & \vdots & 1 \\ 0 & 1 & \vdots & 2 \end{bmatrix}$

13. $\begin{bmatrix} 1 & -2 \\ 1 & 2 \end{bmatrix} \begin{bmatrix} x \\ y \end{bmatrix} = \begin{bmatrix} 1 \\ 3 \end{bmatrix}$

$$\left[\begin{array}{cc|cc} 1 & -2 & 1 & 0 \\ 1 & 2 & 0 & 1 \end{array}\right] \xrightarrow{-①+②} \left[\begin{array}{cc|cc} 1 & -2 & 1 & 0 \\ 0 & 4 & -1 & 1 \end{array}\right] \xrightarrow{\frac{1}{2}②+①}$$

$$\left[\begin{array}{cc|cc} 1 & 0 & \frac{1}{2} & \frac{1}{2} \\ 0 & 4 & -1 & 1 \end{array}\right] \xrightarrow{\frac{1}{4}②} \left[\begin{array}{cc|cc} 1 & 0 & \frac{1}{2} & \frac{1}{2} \\ 0 & 1 & -\frac{1}{4} & \frac{1}{4} \end{array}\right]$$

$$\begin{bmatrix} \frac{1}{2} & \frac{1}{2} \\ -\frac{1}{4} & \frac{1}{4} \end{bmatrix} \begin{bmatrix} 1 \\ 3 \end{bmatrix} = \begin{bmatrix} 2 \\ \frac{1}{2} \end{bmatrix}$$

15. $\left[\begin{array}{ccc|c} 1 & -2 & 3 & 4 \\ -3 & 4 & -1 & -2 \\ 2 & 1 & -4 & 3 \end{array}\right] \xrightarrow{②+③} \left[\begin{array}{ccc|c} 1 & -2 & 3 & 4 \\ -3 & 4 & -1 & -2 \\ -1 & 5 & -5 & 1 \end{array}\right] \xrightarrow{①+③}$

$$\left[\begin{array}{ccc|c} 1 & -2 & 3 & 4 \\ -3 & 4 & -1 & -2 \\ 0 & 3 & -2 & 5 \end{array}\right] \xrightarrow{3①+②} \left[\begin{array}{ccc|c} 1 & -2 & 3 & 4 \\ 0 & -2 & 8 & 10 \\ 0 & 3 & -2 & 5 \end{array}\right] \xrightarrow[②+③]{-②+①}$$

$$\left[\begin{array}{ccc|c} 1 & 0 & -5 & -6 \\ 0 & -2 & 8 & 10 \\ 0 & 1 & 6 & 15 \end{array}\right] \xrightarrow{2③+②} \left[\begin{array}{ccc|c} 1 & 0 & -5 & -6 \\ 0 & 0 & 20 & 40 \\ 0 & 1 & 6 & 15 \end{array}\right] \xrightarrow{\frac{1}{20}②}$$

$$\left[\begin{array}{ccc|c} 1 & 0 & -5 & -6 \\ 0 & 0 & 1 & 2 \\ 0 & 1 & 6 & 15 \end{array}\right] \xrightarrow[-6②+③]{5②+①} \left[\begin{array}{ccc|c} 1 & 0 & 0 & 4 \\ 0 & 0 & 1 & 2 \\ 0 & 1 & 0 & 3 \end{array}\right]$$

17. $y + x = -6$

    $y + x^2 = 0, \quad y = -x^2$

    $-x^2 + x = -6$

    $x^2 - x - 6 = 0$

    $(x-3)(x+2) = 0$

          $x = -2, 3$

          $y = -4, -9$

  $(-2, 4), \quad (3, -9)$

19. $x^2 + y^2 = 6$

   $\underline{x^2 - y^2 = 2}$

       $2x^2 = 8$

    $x^2 = 4, \quad x = \pm 2$

      $y^2 = 6 - x^2$

      $y^2 = 2$

      $y = \pm\sqrt{2}$

  $(\pm 2, \pm\sqrt{2})$

23.
$$\begin{bmatrix} 1 & 1 & 2 \\ 2 & 3 & 2 \\ 0 & 4 & -1 \end{bmatrix} + \begin{bmatrix} -3 & 7 & 6 \\ 0 & -2 & 3 \\ -2 & 7 & 1 \end{bmatrix} = \begin{bmatrix} -2 & 8 & 8 \\ 2 & 1 & 5 \\ -2 & 11 & 0 \end{bmatrix}$$

25. $AB = \begin{bmatrix} 2 & 3 & -1 \end{bmatrix} \begin{bmatrix} -1 \\ 2 \\ 4 \end{bmatrix} = \begin{bmatrix} 2(-1)+3(2)-1(4) \end{bmatrix} = \begin{bmatrix} 0 \end{bmatrix}$

$BA = \begin{bmatrix} -1 \\ 2 \\ 4 \end{bmatrix} \begin{bmatrix} 2 & 3 & -1 \end{bmatrix} = \begin{bmatrix} -1(2) & -1(3) & -1(-1) \\ 2(2) & 2(3) & 2(-1) \\ 4(2) & 4(3) & 4(-1) \end{bmatrix}$

$$= \begin{bmatrix} -2 & -3 & 1 \\ 4 & 6 & -2 \\ 8 & 12 & -4 \end{bmatrix}$$

27. $AB = \begin{bmatrix} 2 & -1 \\ 3 & 0 \end{bmatrix} \begin{bmatrix} 5 & -3 \\ 1 & 7 \end{bmatrix} = \begin{bmatrix} 2(5)-1(1) & 2(-3)-1(7) \\ 3(5)+0(1) & 3(-3)+0(7) \end{bmatrix} = \begin{bmatrix} 9 & -13 \\ 15 & -9 \end{bmatrix}$

$BA = \begin{bmatrix} 5 & -3 \\ 1 & 7 \end{bmatrix} \begin{bmatrix} 2 & -1 \\ 3 & 0 \end{bmatrix} = \begin{bmatrix} 5(2)-3(3) & 5(-1)-3(0) \\ 1(2)+7(3) & 1(-1)+7(0) \end{bmatrix} = \begin{bmatrix} 1 & -5 \\ 12 & -1 \end{bmatrix}$

29. $AB = \begin{bmatrix} 2 & 1 & -1 \\ 4 & 0 & 1 \end{bmatrix} \begin{bmatrix} 2 & -3 \\ 1 & 2 \\ 4 & 8 \end{bmatrix} =$

$$= \begin{bmatrix} 2(2)+1(1)-1(4) & 2(-3)+1(2)-1(8) \\ 4(2)+0(1)+1(4) & 4(-3)+0(2)+1(8) \end{bmatrix} = \begin{bmatrix} 1 & -12 \\ 12 & -4 \end{bmatrix}$$

$BA = \begin{bmatrix} 2 & -3 \\ 1 & 2 \\ 4 & 8 \end{bmatrix} \begin{bmatrix} 2 & 1 & -1 \\ 4 & 0 & 1 \end{bmatrix} =$

$$= \begin{bmatrix} 2(2)-3(4) & 2(1)-3(0) & 2(-1)-3(1) \\ 1(2)+2(4) & 1(1)+2(0) & 1(-1)+2(1) \\ 4(2)+8(4) & 4(1)+8(0) & 4(-1)+8(1) \end{bmatrix} = \begin{bmatrix} -8 & 2 & -5 \\ 10 & 1 & 1 \\ 40 & 4 & 4 \end{bmatrix}$$

31. $\left[\begin{array}{cc|cc} 1 & 2 & 1 & 0 \\ 3 & 2 & 0 & 1 \end{array}\right] \xrightarrow{-3①+②} \left[\begin{array}{cc|cc} 1 & 2 & 1 & 0 \\ 0 & -4 & -3 & 1 \end{array}\right] \xrightarrow{-\frac{1}{4}③}$

$\left[\begin{array}{cc|cc} 1 & 2 & 1 & 0 \\ 0 & 1 & \frac{3}{4} & -\frac{1}{4} \end{array}\right] \xrightarrow{-2②+①} \left[\begin{array}{cc|cc} 1 & 0 & -\frac{1}{2} & \frac{1}{2} \\ 0 & 1 & \frac{3}{4} & -\frac{1}{4} \end{array}\right]$

33. $\left[\begin{array}{cc|cc} -1 & 4 & 1 & 0 \\ 3 & 5 & 0 & 1 \end{array}\right] \xrightarrow{3①+②} \left[\begin{array}{cc|cc} -1 & 4 & 1 & 0 \\ 0 & 17 & 3 & 1 \end{array}\right] \begin{array}{l} \xrightarrow{\frac{1}{17}②} \\ \xrightarrow{-①} \end{array}$

$\left[\begin{array}{cc|cc} 1 & -4 & -1 & 0 \\ 0 & 1 & \frac{3}{17} & \frac{1}{17} \end{array}\right] \xrightarrow{4②+①} \left[\begin{array}{cc|cc} 1 & 0 & -\frac{5}{17} & \frac{4}{17} \\ 0 & 1 & \frac{3}{17} & \frac{1}{17} \end{array}\right]$

35. $\begin{bmatrix} 1 & 1 & -1 & | & 1 & 0 & 0 \\ 2 & 0 & 1 & | & 0 & 1 & 0 \\ 0 & -2 & 3 & | & 0 & 0 & 1 \end{bmatrix} \xrightarrow[\substack{\textcircled{2}}]{-2\textcircled{1}+} \begin{bmatrix} 1 & 1 & -1 & | & 1 & 0 & 0 \\ 0 & -2 & 3 & | & -2 & 1 & 0 \\ 0 & -2 & 3 & | & 0 & 0 & 1 \end{bmatrix} \xrightarrow{-\textcircled{2}+\textcircled{3}}$

$\begin{bmatrix} 1 & 1 & -1 & | & 1 & 0 & 0 \\ 0 & -2 & 3 & | & -2 & 1 & 0 \\ 0 & 0 & 0 & | & -2 & 1 & 1 \end{bmatrix}$ Inverse does not exist.

37. $\begin{bmatrix} 3 & 5 & | & 1 & 0 \\ -1 & 3 & | & 0 & 1 \end{bmatrix} \xrightarrow{3\textcircled{2}+\textcircled{1}} \begin{bmatrix} 0 & 14 & | & 1 & 3 \\ -1 & 3 & | & 0 & 1 \end{bmatrix} \xrightarrow[\frac{1}{14}\textcircled{1}]{-\textcircled{2}}$

$\begin{bmatrix} 0 & 1 & | & \frac{1}{14} & \frac{3}{14} \\ 1 & -3 & | & 0 & 1 \end{bmatrix} \xrightarrow{3\textcircled{1}+\textcircled{2}} \begin{bmatrix} 0 & 1 & | & \frac{1}{14} & \frac{3}{14} \\ 1 & 0 & | & \frac{3}{14} & -\frac{5}{14} \end{bmatrix} \xrightarrow{\textcircled{1}\leftrightarrow\textcircled{2}}$

$\begin{bmatrix} 1 & 0 & | & \frac{3}{14} & -\frac{5}{14} \\ 0 & 1 & | & \frac{1}{14} & \frac{3}{14} \end{bmatrix}$

$\begin{bmatrix} 3 & 5 \\ -1 & 3 \end{bmatrix} \begin{bmatrix} x \\ y \end{bmatrix} = \begin{bmatrix} 13 \\ 5 \end{bmatrix}, \qquad \begin{bmatrix} \frac{3}{14} & -\frac{5}{14} \\ \frac{1}{14} & \frac{3}{14} \end{bmatrix} \begin{bmatrix} 13 \\ 5 \end{bmatrix} = \begin{bmatrix} 1 \\ 2 \end{bmatrix}$

39. $\begin{bmatrix} 1 & -2 & | & 1 & 0 \\ 1 & 2 & | & 0 & 1 \end{bmatrix} \xrightarrow{-\textcircled{1}+\textcircled{2}} \begin{bmatrix} 1 & -2 & | & 1 & 0 \\ 0 & 4 & | & -1 & 1 \end{bmatrix} \xrightarrow{\frac{1}{4}\textcircled{2}}$

$\begin{bmatrix} 1 & -2 & | & 1 & 0 \\ 0 & 1 & | & -\frac{1}{4} & \frac{1}{4} \end{bmatrix} \xrightarrow{2\textcircled{2}+\textcircled{1}} \begin{bmatrix} 1 & 0 & | & \frac{1}{2} & \frac{1}{2} \\ 0 & 1 & | & -\frac{1}{4} & \frac{1}{4} \end{bmatrix}$

$\begin{bmatrix} 1 & -2 \\ 1 & 2 \end{bmatrix} \begin{bmatrix} x \\ y \end{bmatrix} = \begin{bmatrix} 1 \\ 3 \end{bmatrix}, \qquad \begin{bmatrix} \frac{1}{2} & \frac{1}{2} \\ -\frac{1}{4} & \frac{1}{4} \end{bmatrix} \begin{bmatrix} 1 \\ 3 \end{bmatrix} = \begin{bmatrix} 2 \\ \frac{1}{2} \end{bmatrix}$

41.

$$\left[\begin{array}{ccc|ccc} 1 & -2 & 3 & 1 & 0 & 0 \\ -3 & 4 & -1 & 0 & 1 & 0 \\ 2 & 1 & 4 & 0 & 0 & 1 \end{array}\right] \xrightarrow[-2①+③]{3①+②}$$

$$\left[\begin{array}{ccc|ccc} 1 & -2 & 3 & 1 & 0 & 0 \\ 0 & -2 & 8 & 3 & 1 & 0 \\ 0 & 5 & -10 & -2 & 0 & 1 \end{array}\right] \xrightarrow{-\frac{1}{2}②}$$

$$\left[\begin{array}{ccc|ccc} 1 & -2 & 3 & 1 & 0 & 0 \\ 0 & 1 & -4 & -\frac{3}{2} & -\frac{1}{2} & 0 \\ 0 & 5 & -10 & -2 & 0 & 1 \end{array}\right] \xrightarrow[-5②+③]{2②+①}$$

$$\left[\begin{array}{ccc|ccc} 1 & 0 & -5 & -2 & -1 & 0 \\ 0 & 1 & -4 & -\frac{3}{2} & -\frac{1}{2} & 0 \\ 0 & 0 & 10 & \frac{11}{2} & \frac{5}{2} & 1 \end{array}\right] \xrightarrow{\frac{1}{10}③}$$

$$\left[\begin{array}{ccc|ccc} 1 & 0 & -5 & -2 & 1 & 0 \\ 0 & 1 & -4 & -\frac{3}{2} & -\frac{1}{2} & 0 \\ 0 & 0 & 1 & \frac{11}{20} & \frac{1}{4} & \frac{1}{10} \end{array}\right] \xrightarrow[4③+②]{5③+①}$$

$$\left[\begin{array}{ccc|ccc} 1 & 0 & 0 & \frac{15}{20} & \frac{1}{4} & \frac{1}{2} \\ 0 & 1 & 0 & \frac{7}{20} & \frac{1}{2} & \frac{4}{10} \\ 0 & 0 & 1 & \frac{11}{20} & \frac{1}{4} & \frac{1}{10} \end{array}\right]$$

$$\left[\begin{array}{ccc} 1 & -2 & 3 \\ -3 & 4 & -1 \\ 2 & 1 & 4 \end{array}\right]\left[\begin{array}{c} a \\ b \\ c \end{array}\right] = \left[\begin{array}{c} 4 \\ -2 \\ 3 \end{array}\right], \qquad \left[\begin{array}{ccc} \frac{15}{20} & \frac{5}{20} & \frac{10}{20} \\ \frac{14}{20} & \frac{10}{20} & \frac{8}{20} \\ \frac{11}{20} & \frac{5}{20} & \frac{2}{20} \end{array}\right]\left[\begin{array}{c} 4 \\ -2 \\ 3 \end{array}\right] = \left[\begin{array}{c} 4 \\ 3 \\ 2 \end{array}\right]$$

43. $\begin{vmatrix} 2 & -1 \\ 3 & 2 \end{vmatrix} = 4 - (-3) = 7$ **45.** $\begin{vmatrix} -8 & 6 \\ 5 & -2 \end{vmatrix} = 16 - (30) = -14$

47. $\begin{vmatrix} 7 & -3 & 2 \\ 4 & -5 & 12 \\ -21 & 9 & -6 \end{vmatrix} = 210 + 756 + 72 - (210+756+72) = 0$

49.
$$
\begin{vmatrix} 1 & 5 & 4 \\ -2 & 3 & -1 \\ 2 & -1 & 5 \end{vmatrix} = 1 \begin{vmatrix} 3 & -1 \\ -1 & 5 \end{vmatrix} - 5 \begin{vmatrix} -2 & -1 \\ 2 & 5 \end{vmatrix} + 4 \begin{vmatrix} -2 & 3 \\ 2 & -1 \end{vmatrix} =
$$

$$
= 1(15-1) - 5(-10+2) + 4(2-6) = 38
$$

51.
$$
\begin{vmatrix} 2 & 2 & -1 & 1 \\ 0 & 1 & 3 & 2 \\ -1 & 0 & 1 & 1 \\ 0 & 2 & -2 & 1 \end{vmatrix} \xrightarrow{2\,\textcircled{3}+\textcircled{1}} \begin{vmatrix} 0 & 2 & 1 & 3 \\ 0 & 1 & 3 & 2 \\ -1 & 0 & 1 & 1 \\ 0 & 2 & -2 & 1 \end{vmatrix} =
$$

$$
-1 \begin{vmatrix} 2 & 1 & 3 \\ 1 & 3 & 2 \\ 2 & -2 & 1 \end{vmatrix} = -\left[6 + 4 - 6 - (18-8+1) \right] = 7
$$

53.
$$
\begin{vmatrix} 2 & 5 \\ x & 1 \end{vmatrix} = 0
$$

$$
2 - 5x = 0
$$
$$
x = \tfrac{2}{5}
$$

55. $x = \dfrac{\begin{vmatrix} 8 & -2 \\ 19 & -3 \end{vmatrix}}{\begin{vmatrix} 5 & -2 \\ 4 & -3 \end{vmatrix}} = \dfrac{14}{-7} = -2, \quad y = \dfrac{\begin{vmatrix} 5 & 8 \\ 4 & 19 \end{vmatrix}}{-7} = \dfrac{63}{-7} = -9$

57.

$$a = \frac{\begin{vmatrix} 1 & -1 & 3 \\ 3 & 0 & -2 \\ 0 & 0 & 5 \end{vmatrix}}{\begin{vmatrix} 0 & -1 & 3 \\ 1 & 0 & -2 \\ 2 & 0 & 5 \end{vmatrix}} = \frac{1\begin{vmatrix} 3 & -2 \\ 0 & 5 \end{vmatrix}}{1\begin{vmatrix} 1 & -2 \\ 2 & 5 \end{vmatrix}} = \frac{15}{9} = \frac{5}{3}$$

$$b = \frac{\begin{vmatrix} 0 & 1 & 3 \\ 1 & 3 & -2 \\ 2 & 0 & 5 \end{vmatrix}}{9} = \frac{-1\begin{vmatrix} 1 & -2 \\ 2 & 5 \end{vmatrix} + 3\begin{vmatrix} 1 & 3 \\ 2 & 0 \end{vmatrix}}{9} = \frac{-5 - 22}{2} = -3$$

$$c = \frac{\begin{vmatrix} 0 & -1 & 1 \\ 1 & 0 & 3 \\ 2 & 0 & 0 \end{vmatrix}}{9} = \frac{1\begin{vmatrix} 1 & 3 \\ 2 & 0 \end{vmatrix}}{9} = \frac{-6}{9} = \frac{-2}{3}$$

59.

$$r = \frac{\begin{vmatrix} 4 & -1 & -1 \\ 2 & 3 & -2 \\ 4 & 6 & -4 \end{vmatrix}}{\begin{vmatrix} 3 & -1 & -1 \\ 1 & 3 & -2 \\ 2 & 6 & -4 \end{vmatrix}} = \frac{48 + 8 - 12 - (-12 - 48 - 8)}{-36 + 4 - 6 - (-6 - 36 + 4)} = \frac{112}{0}, \quad \text{Dependent}$$

61. $x + y = 90$
$x + 3 = y$

$x + y = 90$
$x - y = -3$

$$2x = 87, \quad x = 43.5$$
$$y = 90 - 43.5 = 46.5$$

63.
$$\begin{aligned} x + \quad y &= \quad 750,000 \\ .09x + .11y &= \quad 80,000 \end{aligned}$$

$$\begin{aligned} -.09x - .09y &= -\ 67,000 \\ .09x + .11y &= \quad 80,000 \end{aligned}$$

$$\begin{aligned} .02y &= \quad 12,500 \\ y &= \quad 625,000 \text{ at } 9\% \\ x &= \quad 125,000 \text{ at } 11\% \end{aligned}$$

$x \to 10\%, \quad y \to 30\%, \quad z \to 50\%$

65. $x + y + z = 250$

$\qquad 2y = z$

$.10x + .30y + .50z = .40(250)$

$x + y + 2y = 250$ $10x + \quad 30y = \quad 2,500$

$10x + 30y + 50(2y) = 10,000$ $10x + 130y = 10,000$

$\qquad\qquad\qquad\qquad\qquad\qquad 100y = \quad 7,500$

$\qquad\qquad\qquad\qquad\qquad\qquad y = 75, \quad z = 150$

$\qquad\qquad\qquad\qquad\qquad\qquad x = 250-75-150 = 25$

71. $x + y = 12$

$\qquad x = 8, \quad y = 4$

$p = x + 2y$

(0, 12) $p = 0+2(12) = 24$, maximum

(8, 0) $p = 8+2(0) = 8$

(8, 4) $p = 8+2(4) = 16$

(0, 0) $p = 0$

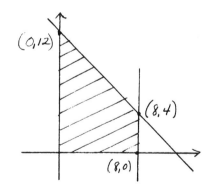

73. $x + \quad y = 4$

$\quad x + 2y = 6$

$\qquad -y = -2, \quad y = 2, \quad x = 2$

$p = 3x + 5y$

(0, 4) $p = 3(0)+5(4) = 20$

(6, 0) $p = 3(6)+5(0) = 18$

(2, 2) $p = 3(2)+5(2) = 16$, minimum

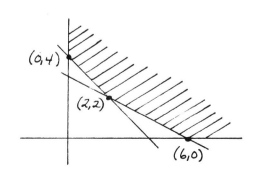

Chapter 11 Vectors and Complex Numbers

1. $|F| = \sqrt{1^2+2^2} = \sqrt{5} = 2.24$

    $\tan \theta = 2, \ \theta = 63.4°$

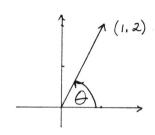

3. $|F| = \sqrt{2+7} = 3$

    $\tan \theta = \dfrac{\sqrt{7}}{\sqrt{2}}, \ \theta = 61.9°$

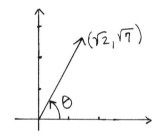

5. $|F| = \sqrt{(-4)^2+(-4)^2} = \sqrt{32} = 5.66$

    $\tan \theta = 1, \ \theta = 135°$

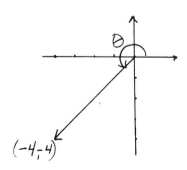

7. $|\mathbf{F}| = \sqrt{4^2+(-3)^2} = \sqrt{25} = 5$

$\tan\theta = \frac{-3}{4}, \theta = -36.9°$

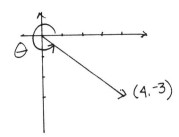

9. $|\mathbf{F}| = \sqrt{3+6^2} = \sqrt{39} = 6.24$

$\tan\theta = \frac{6}{-\sqrt{3}}, \theta = 106.1°$

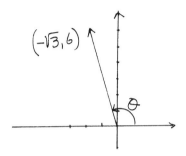

11. $F_x = 10\cos 50° = 6.43$

$F_y = 10\sin 50° = 7.66$

13. $F_x = 13.7\cos 34°10' = 11.34$

$F_y = 13.7\sin 34°10' = 7.69$

15. $F_x = 158\cos 125° = -90.63$

$F_y = 158\sin 125° = 129.43$

17. $F_x = 43.5\cos 220° = -33.32$

$F_y = 43.5\sin 220° = -27.96$

19. $F_x = 10.4\cos 335° = 9.43$

$F_y = 10.4\sin 335° = -4.40$

21. $|\mathbf{F}| = \sqrt{20^2+15^2} = 25$

$\tan\theta = \frac{15}{20}, \theta = 36.87°$

23. $|\mathbf{F}| = \sqrt{17.5^2+69.3^2} = 71.5$

$\tan\theta = \frac{69.3}{17.5}, \theta = 75.83°$

25. $|\mathbf{F}| = \sqrt{.130^2+.080^2} = .153$

$\tan\theta = \frac{.080}{.130}, \theta = 31.6°$

27. $|\mathbf{F}| = \sqrt{64+49} = 10.6$

$\tan\theta = \frac{-7}{8}, \theta = 318.8°$

29. $|\mathbf{F}| = \sqrt{4+49} = 7.28$

$\tan\theta = \frac{-7}{-2}, \theta = 254.1°$

31. $F_y = 75\sin 12°10' = 15.8$, vertical

$F_x = 75\cos 12°10' = 73.3$, horizontal

33. $|\mathbf{V}| = \sqrt{300^2+900^2} = 948.7$

35. $|F| = \sqrt{3^2+5^2} = 5.83$

$\tan\theta = \frac{5}{3}$, $\theta = 59°04'$ with shore line

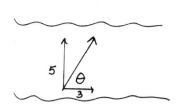

Section 11.2   Operations on Vectors

1. $A+B = (2+3)i + (2-1)j$
$\quad\quad = 5i+j$

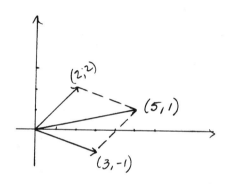

3. $A+B = (-2-4)i + (3-5)j = -6i - 2j$

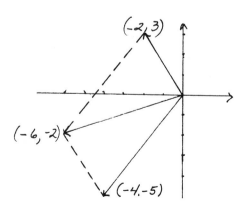

5. $A+B = (-1 + 2)i + (-1 - 2)j = +i - 3j$

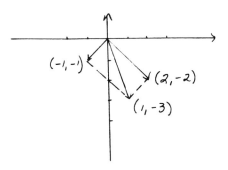

7. $A+B = (-4 + 4)i + (-6 + 2)j = -4j$

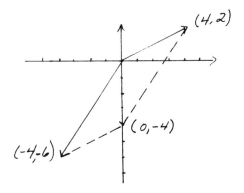

9. $A+B = (5+1)i + (-5 - 2)j = 6i-7j$

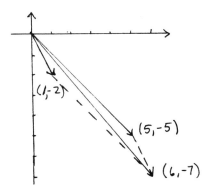

11. $A_x = 16 \cos 25° = 14.50$ $A_y = 16 \sin 25° = 6.76$

     $B_x = 22 \cos 70° = 7.52$ $B_y = 22 \sin 70° = 20.67$

     $R_x = 14.5 + 7.52 = 22.02$ $R_y = 6.76 + 20.67 = 27.43$

     $|\mathbf{R}| = \sqrt{22.02^2 + 27.43^2} = 35.18$

     $\tan \theta = \dfrac{27.43}{22.02}, \; \theta = 51.2°$

13. $A_x = 9.5 \cos 90° = 0$ $A_y = 9.5 \sin 90° = 9.5$

     $B_x = 5.1 \cos 40° = 3.91$ $B_y = 5.1 \sin 40° = 3.28$

     $R_x = 0 + 3.91 = 3.91$ $R_y = 9.5 + 3.28 = 12.78$

     $|\mathbf{R}| = \sqrt{3.91^2 + 12.78^2} = 13.36$

     $\tan \theta = \dfrac{12.78}{3.91}, \; \theta = 72.99°$

15. $A_x = 29.2 \cos 15.6° = 28.12$ $A_y = 29.2 \sin 15.6° = 7.85$

     $B_x = 82.6 \cos 150° = -71.53$ $B_y = 82.6 \sin 150° = 41.3$

     $R_x = 28.12 - 71.53 = -43.41$ $R_y = 7.85 + 41.3 = 49.15$

     $|\mathbf{R}| = \sqrt{43.41^2 + 49.15^2} = 65.58$

     $\tan \theta = \dfrac{49.15}{-43.41}, \; \theta = 131.45°$

17. $A_x = 550 \cos 140° = -421.32$ $A_y = 550 \sin 140° = 353.53$

     $B_x = 925 \cos 310° = 594.58$ $B_y = 925 \sin 310° = -708.59$

     $R_x = -421.32 + 594.58 = 173.26$ $R_y = 353.53 - 708.59 = -355.06$

     $|\mathbf{R}| = \sqrt{173.26^2 + 355.06^2} = 395.08$

     $\tan \theta = \dfrac{-355.06}{173.26}, \; \theta = -63.99°$

19. $\tan \theta = \dfrac{300}{50}, \; \theta = 99.46°$

     $|\mathbf{R}| = \sqrt{300^3 + 50^2} = 304.14$

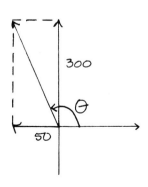

21. $R_x = 200 \cos 35° = 163.83$

23. $A_x = 60$ $A_y = 0$

     $B_x = 75 \cos 55° = 43.02$ $B_y = 75 \sin 55° = 61.44$

     $R_x = 103.02$ $R_y = 61.44$

     $|R| = \sqrt{103.02^2 + 61.44^2} = 119.94$

     $\tan \theta = \dfrac{61.44}{103.02}, \theta = 30.8°$

25. $|B| = 1200, \theta_B = -20°$

     $|P| = 500, \theta_P = 0°$

     $B_x = 1200 \cos (-20°) = 1127.63$ $B_y = 1200 \sin (-20°) = -410.42$

     $P_x = 500 \cos 0° = 500$ $P_y = 500 \sin 0° = 0$

     $R_x = 1627.63$ $R_y = -410.42$

     $|R| = \sqrt{1627.63^2 + 410.42^2} = 1679$

     $\tan \theta = \dfrac{-410.42}{1627.63}, \theta = -14.2°$

Section 11.3   The Dot Product of Two Vectors

1. $\mathbf{A} \cdot \mathbf{B} = 1(2) + 1(-1) = 1$ 3. $\mathbf{A} \cdot \mathbf{B} = 6(7) + 1(-1) = 41$

5. $\mathbf{A} \cdot \mathbf{B} = 1(1) + 0(3) = 1$ 7. $\mathbf{A} \cdot \mathbf{B} = 25(40) \cos (85° - 27°) = 529.9$

9. $\mathbf{A} \cdot \mathbf{B} = 190(75) \cos{(205° - 100°)} = -3688$

11. $\mathbf{A} \cdot \mathbf{B} = (24.3)(16.2) \cos{(276° - 245°)} = 337.4$

13. $\mathbf{A} \cdot \mathbf{B} = 10 - 1 = 9, |\mathbf{A}| = \sqrt{5^2 + 1} = \sqrt{26}, |\mathbf{B}| = \sqrt{2^2 + 1} = \sqrt{5}$
$\cos{\theta} = \dfrac{\mathbf{A} \cdot \mathbf{B}}{|\mathbf{A}||\mathbf{B}|} = \dfrac{9}{\sqrt{5}\sqrt{26}}, \theta = 37.87°$

15. $\mathbf{A} \cdot \mathbf{B} = -6 - 12 = -18, |\mathbf{A}| = \sqrt{9 + 16} = 5, |\mathbf{B}| = \sqrt{4 + 9} = \sqrt{13}$
$\cos{\theta} = \dfrac{-18}{5\sqrt{13}}, \theta = 176.82°$

17. $W = |20||30| \cos{28°} = 529.8$ 19. $W = |5||100| \cos{30°} = 433$

Section 11.4   Complex Numbers

7. $i^{101} = i^{4(25)+1} = (i^4)^{25} i = i$

9. $i^{402} = i^{4(100)+2} = (i^4)^{100} i^2 = i^2 = -i$

17. $x^2 + iy = 9 - 3i$ 19. $3x - y + 7 + i(x + y - 1) = 0$
$(x^2 - 9) + (y + 3)i = 0$ $3x - y + 7 = 0$
$x^2 - 9 = 0 \qquad y = -3$ $\underline{x + y - 1 = 0}$
$x = \pm 3$ $4x \qquad + 6 = 0$
$x = -\dfrac{3}{2}$
$y = 1 - x = \dfrac{5}{2}$

21. $(3 + 2i) + (4 + 3i) = (3 + 4) + (2 + 3)i = 7 + 5i$

23. $(5 - 2i) + (-7 + 5i) = (5 - 7) + (-2 + 5i) = -2 + 3i$

25. $(1 + i) + (3 - i) = (1 + 3) + (1 - 1)i = 4$

27. $(3 + 5i) - 4i = (3 + 0) + (5 - 4)i = 3 + i$

29. $(2+3i)(4+5i) = 8 + 10i + 12i + 15i^2 = -7 + 22i$

31. $(5-i)(5+i) = 25-i^2 = 26$

33. $(4+\sqrt{3}i)^2 = 16 + 8\sqrt{3}i + 3i^2 = 13 + 8\sqrt{3}i$

35. $6i(4-3i) = 24i - 18i^2 = 18+24i$

37. $\dfrac{3+2i}{1+i} \cdot \dfrac{1-i}{1-i} = \dfrac{3-3i+2i-2i^2}{1-i^2} = \dfrac{5-i}{2}$

39. $\dfrac{3}{2-3i} \cdot \dfrac{2+3i}{2+3i} = \dfrac{6+9i}{4-9i^2} = \dfrac{6+9i}{13}$

41. $\dfrac{1}{5i} \cdot \dfrac{-5i}{-5i} = \dfrac{-5i}{-25i^2} = \dfrac{-5i}{25} = \dfrac{-i}{5}$

43. $\dfrac{-1-3i}{4-\sqrt{2}i} \cdot \dfrac{4+\sqrt{2}i}{4+\sqrt{2}i} = \dfrac{-4-\sqrt{2}i-12i-3\sqrt{2}i^2}{16-2i^2} = \dfrac{(-4+3\sqrt{2})-(12+\sqrt{2}i)}{18}$

45. $(a+bi)(a-bi) = a^2-bi^2 = a^2+b^2$

47. $a+bi = \overline{a+bi}$
$a+bi = a-bi$
$0+2bi = 0$
$\quad 2b = 0,\ b = 0$

49. $\dfrac{z+\bar{z}}{2} = \dfrac{(x+iy)+(x-iy)}{2} = \dfrac{2x}{2} = x$

$\dfrac{z-\bar{z}}{2i} = \dfrac{(x+iy)-(x-iy)}{2i} = \dfrac{2iy}{2i} = y$

1. $(4+i) + (3+5i) = 7+6i$

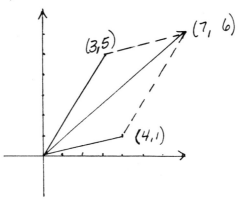

3. $(4+3i) + (-2 + i) = 2+4i$

5. $(2-4i) + (-3 + i) = -1 - 3i$

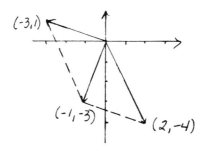

7. $(5+3i) - 6 = -1 + 3i$

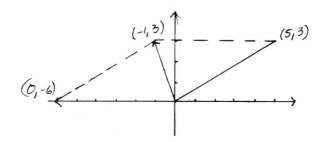

9. $(5-3i) + (5+3i) = 10$

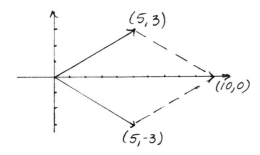

11. $(1+3i) - (2-5i) = -1 + 8i$

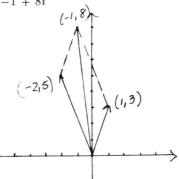

13. $(2+\sqrt{3}i) - (-1 - i) = 3 + (\sqrt{3}+i)i$

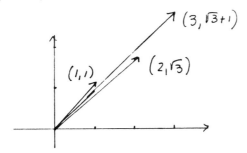

21. $r = \sqrt{1+3} = 2$, $\tan\theta = -\sqrt{3}$, $\theta = -60°$

$z = 2 \text{ cis } (-60°)$

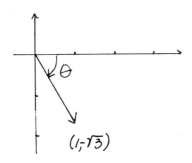

23. $r = \sqrt{5+4} = 3$, $\tan \theta = \frac{2}{\sqrt{5}}$, $\theta = 41.8°$

$z = 3 \text{ cis } 41.8°$

27. $r = \sqrt{9+16} = 5$; $\tan\theta = \frac{-4}{3}$, $\theta = -53.1°$

$z = \text{cis } (-53.1°)$

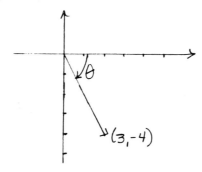

29. $r = \sqrt{25+36} = \sqrt{61}$, $\tan \theta = \frac{-6}{5}$, $\theta = -50.2°$

$z = \sqrt{61} \text{ cis } (-50.2°)$

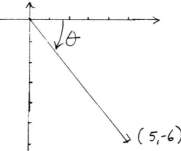

31. $2 \text{ cis } 30° = 2 \cos 30° + 2i \sin 30°$

$= 2\frac{\sqrt{3}}{2} + i2 \cdot \frac{1}{2} = \sqrt{3} + i$

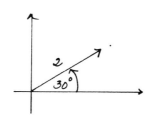

33. $5 \text{ cis } 135° = 5 \cos 135° + 5i \sin 135°$

$$= 5\left(-\frac{\sqrt{2}}{2}\right) + i5\left(\frac{\sqrt{2}}{2}\right) = \frac{5}{2}\sqrt{2}(-1 + i)$$

35. $\sqrt{3} \text{ cis } 210° = \sqrt{3} \cos 210° + \sqrt{3}i \sin 210°$

$$= \sqrt{3}\left(-\frac{\sqrt{3}}{2}\right) + \sqrt{3}\left(-\frac{1}{2}\right)i = \frac{-3}{2} - \frac{\sqrt{3}}{2}i$$

37. $3 \text{ cis } 300° = 3 \cos 300° + 3i \sin 300°$

$$= 3\left(\frac{1}{2}\right) + i(3)\left(\frac{-\sqrt{3}}{2}\right) = \frac{3}{2} - \frac{3\sqrt{3}}{2}i$$

39. $10 \text{ cis } 20° = 10 \cos 20° + 10i \sin 20°$

$$= 9.397 + 3.420i$$

41. $(4 \text{ cis } 30°)(3 \text{ cis } 60°) = 3 \bullet 4 \text{ cis } (30°+60°) = 12 \text{ cis } 90°$

43. $(\sqrt{2} \text{ cis } 90°)(\sqrt{2} \text{ cis } 240°) = (\sqrt{2})^2 \text{ cis } (90° + 240°) = 2 \text{ cis } 330°$

45. $(10 \text{ cis } 35°)(2 \text{ cis } 100°) = 10 \cdot 2 \text{ cis } (35° + 100°) = 20 \text{ cis } 135°$

47. $3 + 4i, \quad r = \sqrt{9 + 16} = 5, \tan \theta = \frac{4}{3}, \theta = 53.1°$

$$3 + 4i = 5 \text{ cis } 53.1°$$

$$\sqrt{3} - i, \quad r = \sqrt{3 + 1} = 2, \tan \theta = \frac{-1}{\sqrt{3}}, \theta = -30°$$

$$\sqrt{3} - i = 2 \text{ cis } (-30°)$$

$$(3 + 4i)(\sqrt{3} - i) = (5 \text{ cis } 53.1°)\left(2 \text{ cis } (-30°)\right) = 10 \text{ cis } 23.1°$$

49. $\dfrac{10 \text{ cis } 30°}{2 \text{ cis } 90°} = \dfrac{10}{2} \text{ cis } (30° - 90°) = 5 \text{ cis } (-60°)$

51. $\dfrac{4 \text{ cis } 26°40'}{2 \text{ cis } 19°10'} = \dfrac{4}{2} \text{ cis } (26°40' - 19°10') = 2 \text{ cis } 7°30'$

53. $1 - i, \quad r = \sqrt{1 + 1} = \sqrt{2}, \tan \theta = -1, \theta = -45°$

$$1 - i = \sqrt{2} \text{ cis } (-45°)$$

$$\sqrt{3} + i, \quad r = \sqrt{3 + 1} = 2, \tan \theta = \frac{-1}{\sqrt{3}}, \theta = 30°$$

$$\sqrt{3} + i = 2 \text{ cis } 30°$$

$$\frac{1 - i}{\sqrt{3} + i} = \frac{\sqrt{2} \text{ cis } (-45°)}{2 \text{ cis } 30°} = \frac{\sqrt{2}}{2} \text{ cis } (-45° - 30°) = \frac{\sqrt{2}}{2} \text{ cis } (-75°)$$

55. $4i, \qquad r = 4, \theta = 90°, 4i = 4 \text{ cis } 90°$

$$-1 + i, \quad r = \sqrt{1 + 1} = \sqrt{2}, \tan \theta = -1, \theta = 135°$$

$$-1 + i = \sqrt{2} \text{ cis } 135°$$

$$\frac{4i}{-1 + i} = \frac{4 \text{ cis } 90°}{\sqrt{2} \text{ cis } 135°} = \frac{4}{\sqrt{2}} \text{ cis } (90° - 135°) = \frac{4}{\sqrt{2}} \text{ cis } (-45°)$$

57. $\cos \theta$

$$\tfrac{1}{2}\left(\cos \theta + \text{ cis } (-\theta)\right) = \tfrac{1}{2}\left(\cos \theta + i \sin \theta + \cos (-\theta) + i \sin (-\theta)\right)$$

$$= \tfrac{1}{2}(\cos \theta + i \sin \theta + \cos \theta - i \sin \theta)$$

$$= \tfrac{1}{2}(2 \cos \theta) = \cos \theta$$

$$\tfrac{1}{2i}\left(\text{ cis } \theta - \text{ cis } (-\theta)\right) = \tfrac{1}{2i}\left(\cos \theta + i \sin \theta - \cos (-\theta) - i \sin (-\theta)\right)$$

$$= \tfrac{1}{2i}(\cos \theta + i \sin \theta - \cos \theta + i \sin \theta)$$

$$= \tfrac{1}{2i}(2i \sin \theta) = \sin \theta$$

1. $(-1 + \sqrt{3}i)^3$, $r = \sqrt{1+3} = 2$, $\tan \theta = -\sqrt{3}$, $\theta = 120°$

   $(-1 + \sqrt{3}i)^3 = \left(2 \text{ cis } (120°)\right)^3 = 2^3 \text{ cis } (360°) = 8 \text{ cis } 0°$

3. $(\sqrt{3} \text{ cis } 60°)^4 = (\sqrt{3})^4 \text{ cis } 4(60°) = 9 \text{ cis } 240°$

5. $(-2 + 2i)^5$, $r = \sqrt{4+4} = \sqrt{8}$, $\tan \theta = \frac{2}{-2} = -1$, $\theta = 135°$

   $(-2 + 2i)^5 = (\sqrt{8} \text{ cis } 135°)^5 = (\sqrt{8})^5 \text{ cis } 5(135°) = 128\sqrt{2} \text{ cis } 675°$

   $= 128\sqrt{2} \text{ cis } 315°$

7. $(-\sqrt{3} + i)^7$, $r = \sqrt{3+1} = 2$, $\tan \theta = \frac{-1}{\sqrt{3}}$, $\theta = 150°$

   $(-\sqrt{3} + i)^7 = \left(2 \text{ cis } (150°)\right)^7 = 2^7 \text{ cis } 7 (150°) = 128 \text{ cis } 1050°$

   $= 128 \text{ cis } 330°$

9. $(2+5i)^4$, $r = \sqrt{4+25} = \sqrt{29}$, $\tan \theta = \frac{5}{2}$, $\theta = 68.2°$

   $(2+5i)^4 = (\sqrt{29} \text{ cis } 68.2°)^4 = 29^2 \text{ cis } 4(68.2°) = 841 \text{ cis } 272.8°$

11. $(1 \text{ cis } 0°)^{1/5} = 1^{1/5} \text{ cis } \left(\frac{0 + k360°}{5}\right)$

   $= \text{cis } 0°, \text{ cis } 72°, \text{ cis } 144°, \text{ cis } 216°, \text{ cis } 288°$

13. $i^{1/4} = (1 \text{ cis } 90°)^{1/4} = \text{cis } \frac{(90° + k360°)}{4}$

   $= \text{cis } 22.5°, \text{ cis } 112.5°, \text{ cis } 202.5°, \text{ cis } 292.5°$

15. $(1+i)^{1/2} = (\sqrt{2} \text{ cis } 45°)^{1/2} = 2^{1/4} \text{ cis } \frac{(45° + k360°)}{2}$

   $= \sqrt[4]{2} \text{ cis } 22.5°, \sqrt[4]{2} \text{ cis } 202.5°$

17. $(-\sqrt{3} + i)^{1/6}$, $r = \sqrt{3+1} = 2$, $\tan \theta = \frac{-1}{\sqrt{3}}$, $\theta = 150°$

$$(-\sqrt{3} + i)^{1/6} = (2 \text{ cis } 150°)^{1/6} = 2^{1/6} \text{ cis } \frac{(150 + k360°)}{6}$$

$$= \sqrt[6]{2} \text{ cis } (250° + M60°), \ M = 0, 1, 2, 3, 4, 5, 6$$

19. $(-1 + 3i)^{1/4}$, $r = \sqrt{1+3} = 2$, $\tan \theta = -\sqrt{3}$, $\theta = 120°$

$$(-1 + 3i)^{1/4} = (2 \text{ cis } 120°)^{1/4} = 2^{1/4} \text{ cis } \frac{(120° + k360°)}{4}$$

$$= 2^{1/4} \text{ cis } 30°, \ 2^{1/4} \text{ cis } 120°, \ 2^{1/4} \text{ cis } 210° \ 2^{1/4} \text{ cis } 300°$$

21. $(\text{cis } \theta)^2 = 1 \cos 2\theta$

$$\sin 2\theta = \sqrt{1 - \cos^2 2\theta} = \sqrt{1 - \left((\text{cis } \theta)^2\right)^2} = \sqrt{1 - (\text{cis } \theta)^4}$$

23. $x = (-64)^{1/3}$ $r = 64$, $\theta = 180°$, $-64 = 64 \text{ cis } 180°$

$$(-64)^{1/3} = (64 \text{ cis } 180°)^{1/3} = 64^{1/3} \text{ cis } \frac{(180° + k360°)}{3}$$

$$= 4 \text{ cis } 60°, \ 4 \text{ cis } 180°, \ 4 \text{ cis } 300°$$

Chapter 11   Review

1. $|F| = \sqrt{9+25} = \sqrt{34}$, $\tan \theta = \frac{5}{-3}$, $\theta = 120.96°$

3. $|F| = \sqrt{\pi^2 + (5.4)^2} = 6.25$, $\tan \theta = \frac{-5.4}{\pi}$, $\theta = -59.81°$

5. $|F| = \sqrt{\pi + 2} = 2.268$, $\tan \theta = \frac{-\sqrt{\pi}}{\sqrt{2}}$, $\theta = -51.4°$

7. $x = 15.6 \cos (-134°) = -10.84$
   $y = 15.6 \sin (-134°) = -11.22$

9. $x = 1000 \cos 2.65 = -881.58$

    $y = 1000 \sin 2.65 = 472.03$

11. $A+B = (8i + 4j) + (-2i + j) = (8-2)i + (4+1)j = 6i + 5j$

13. $(i - 3j) + (2i + 3j) = (1 + 2)i + (-3 + 3)j = 3i$

15. $(-i - 2j) + (i - 2j) = (-1 + 1)i + (-2 - 2)j = -4j$

17. $A_x = 100 \cos 30° = 86.6$ $\qquad$ $B_x = 83.2 \cos 41.5° = 62.3$

    $A_y = 100 \sin 30° = 50$ $\qquad$ $B_y = 83.2 \sin 41.5° = 55.1$

    $A_x + B_x = 86.6 + 62.3 = 148.9$

    $A_y + B_y = 50 + 55.1 = 105.1$

    $A+B = 148.9i + 105.1j$

17. $A_x = .5 \cos (-.34) = .471$ $\qquad$ $B_x = 1.3 \cos 1.6 = -.038$

    $A_y = .5 \sin (-.34) = -.167$ $\qquad$ $B_y = 1.3 \sin (1.6) = 1.299$

    $A_x + B_x = .471 - .038 = .433$

    $A_y + B_y = -.167 + 1.299 = 1.133$

    $A+B = .471i + 1.133j$

21. $A \cdot B = 8(-2) + 4(1) = -12$

    $|A| = \sqrt{8^2 + 4^2} = \sqrt{80},\ |B| = \sqrt{2^2 + 1^2} = \sqrt{5}$

    $\cos \theta = \dfrac{-12}{\sqrt{5}\sqrt{80}},\ \theta = 126.9$

23. $A \cdot B = 1(2) - 3(3) = -7$

    $|A| = \sqrt{1^2 + 3^2} = \sqrt{10},\ |B| = \sqrt{2^2 + 3^2} = \sqrt{13}$

    $\cos \theta = \dfrac{-7}{\sqrt{10}\sqrt{13}},\ \theta = 127.9°$

25. $\mathbf{A} \cdot \mathbf{B} = 1(-1) + (-2)(-2) = 3$

   $|\mathbf{A}| = \sqrt{1^2 + 2^2} = \sqrt{5}, |\mathbf{B}| = \sqrt{1^2 + 2^2} = \sqrt{5}$

   $\cos \theta = \frac{3}{\sqrt{5}\sqrt{5}} = \frac{3}{5}, \theta = 53.1°$

27. $\mathbf{A} \cdot \mathbf{B} = (100)(83.2) \cos (41.5° - 30°) = 8320 \cos (11.5°) = 8152.9$

29. $\mathbf{A} \cdot \mathbf{B} = (.5)(1.3) \cos (1.6 + 3.4) = .065 \cos 1.94 = -.235$

   $1.94 = 111.2°$

31. $(3+4i) - (5-2i) = (3-5) + (4+2)i = -2 + 6i$

   $r = \sqrt{2^2 + 6^2} = \sqrt{40}, \tan \theta = \frac{6}{-2}, \theta = 108.4°$

   $-2 + 6i = \sqrt{40} \text{ cis } 108.4°$

33. $(6+3i) - (\overline{6+3i}) = (6+3i) - (6-3i) = 6i$

   $r = 6, \theta = 90°$

   $6i = 6 \text{ cis } 90°$

35. $(2+i)(i-1) = 2i - 2 + i^2 - i = -3 + i$

   $r = \sqrt{3^2+1^2} = \sqrt{10}, \tan \theta = \frac{-1}{3}, \theta = 161.6°$

   $-3 + i = \sqrt{10} \text{ cis } 161.6°$

37. $(2+3i)(4+i) = 8 + 2i + 12i + 3i^2 = 5 + 14i$

   $r = \sqrt{5^2 + 14^2} = \sqrt{221}, \tan \theta = \frac{14}{5}, \theta = 70.3°$

   $5 + 14i = \sqrt{221} \text{ cis } 70.3°$

39.
$$\text{cis } .45 = \cos .45 + i \sin .45 = .8984 + .4350i$$
$$2 \text{ cis } (-.95) = 2 \cos (-.95) + 2i \sin (-.95) = 1.163 - 1.627i$$
$$\text{cis } .45 + 2 \text{ cis } (-.95) = (.8984 + 1.163) + (.4350 - 1.627)i$$
$$= 2.061 - 1.192i$$

   $r = \sqrt{2.061^2 + 1.192^2} = 2.38$

   $\tan \theta = \frac{-1.192}{2.061}, \theta = -30.1°$

   $2.061 - 1.192i = 2.38 \text{ cis } (-30.1°)$

41. $\dfrac{4 \cos 4.6}{\text{cis}\,(-2.9)} = 4 \text{ cis } (4.6 + 2.9) = 4 \text{ cis } 7.5 = 4 \text{ cis } 69.72°$

$\qquad = 4(\cos 69.72° + i \sin 69.72°) = 1.39 + 3.75i$

43. $\dfrac{2+3i}{4+i} \cdot \dfrac{4-i}{4-i} = \dfrac{8-2i+12i-3i^2}{16-i^2} = \dfrac{11}{17} + \dfrac{10i}{17}$

$\qquad r = \sqrt{\left(\dfrac{11}{17}\right)^2 + \left(\dfrac{10}{17}\right)^2} = \dfrac{\sqrt{221}}{17}, \tan \theta = \dfrac{\frac{10}{17}}{\frac{11}{17}} = \dfrac{10}{11}, \theta = 42.3°$

$\qquad \dfrac{11}{17} + \dfrac{10i}{17} = \dfrac{\sqrt{221}}{17} \text{ cis } 42.3°$

45. $(\sqrt{5} \text{ cis } 120°)(7 \text{ cis } 80°) = 7\sqrt{5} \text{ cis } (120° + 80°) = 7\sqrt{5} \text{ cis } 200°$

$\qquad 7\sqrt{5} \, (\cos 200° + i \sin 200°) = -14.7 - 5.35i$

47. $100 \left[\text{cis } (-120\pi) + \text{cis } (120\pi) \right]$

$\qquad = 100 \left[\cos (-120\pi) + i \sin (-120\pi) + \cos (120\pi) + i \sin (120\pi) \right]$

$\qquad = 100 \left[\cos (-120\pi) - i \sin (120\pi) + \cos (120\pi) + i \sin (120\pi) \right]$

$\qquad = 200 \cos 120\pi = 200 \cos 0° = 200$

49. $3.4 + 4i, r = \sqrt{3.4^2 + 4^2} = 27.56, \tan \theta = \dfrac{4}{3.4}, \theta = 49.6°$

$\qquad 3.4 + 4i, r = 27.56, \tan \theta = \dfrac{-4}{3.4}, \theta = -49.6$

$\qquad \dfrac{3.4 + 4i}{3.4 - 4i} = \dfrac{27.56 \text{ cis } 49.6°}{27.56 \text{ cis } (-49.6°)} = \text{cis } 99.2°$

$\qquad\qquad = \cos 99.2° + i \sin 99.2°$

$\qquad\qquad = -.16 + .99i$

51. $\qquad -1 = 1 \text{ cis } 180°$

$\qquad (-1)^{1/3} = (\text{cis } 180°)^{1/3} = \text{cis } \dfrac{(180° + k360°)}{3} = \text{cis } 60°, \text{cis } 180°, \text{cis } 300°$

53. $\qquad i = 1 \text{ cis } 90°$

$\qquad i^{1/3} = (\text{cis } 90°)^{1/3} = \text{cis } \left(\dfrac{90° + k360°}{3}\right) = \text{cis } 30°, \text{cis } 150°, \text{cis } 270°$

55. $i-1$, $r = \sqrt{1+1} = \sqrt{2}$, $\tan \theta = -1$, $\theta = 135°$

$$(i-1)^{1/3} = (\sqrt{2} \text{ cis } 135°)^{1/3} = 2^{1/6} \text{ cis } \frac{(135° + k360°)}{3}$$

$$= 2^{1/6} \text{ cis } 45°, \ 2^{1/6} \text{ cis } 165°, \ 2^{1/6} \text{ cis } 285°$$

57. $1 + 3i$, $r = \sqrt{1+3^2} = \sqrt{10}$, $\tan \theta = \frac{3}{1}$, $\theta = 71.6°$

$$(1 + 3i)^5 = (\sqrt{10} \text{ cis } 71.6°)^5 = 10^{5/2} \text{ cis } 5(71.6)° = 10^{5/2} \text{ cis } 357.8°$$

59. $|F| = \sqrt{3.4^2 + 2.5^2} = 4.22$

$\tan \theta = \frac{-2.5}{3.4}$, $\theta = -36.3°$

61. $|A| = \sqrt{5^2 + 3^2} = \sqrt{34}$, $|B| = \sqrt{1^2 + 3^2} = \sqrt{10}$

$A + B = -5 - 9 = -14$

$\cos \theta = \frac{-14}{\sqrt{34}\sqrt{10}}$, $\theta = 139.4°$

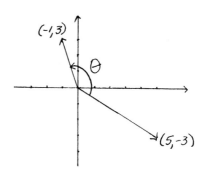

63a. $t^2 + t - 6 = 0$

$(t+3)(t-2) = 0$, $t = 2, -3$

63b. $t^2 - 3t + 2 = 0$

$(t-2)(t-1) = 0$, $t = 1, 2$

63c. $t = 2$

65. $5-4i$, $r = \sqrt{25+16} = \sqrt{41}$, $\tan \theta = \frac{-4}{5}$, $\theta = -38.7°$

$3-i$, $r = \sqrt{9+1} = \sqrt{10}$, $\tan \theta = \frac{-1}{3}$, $\theta = -18.4°$

$$\frac{5-4i}{3-i} = \frac{\sqrt{4} \text{ cis } (-38.7°)}{\sqrt{10} \text{ cis } (-18.4°)} = 2.02 \text{ cis } (-38.7° + 18.4°) = 2.02 \text{ cis } (-20.3°)$$

$$= 2.02 \left(\cos (-20.3°) + i \sin (-20.3°) \right)$$

$$= 1.9 - .7i$$

Chapter 12　The Algebra of Polynomials

The Remainder and Factor Theorems

1.
$$
\begin{array}{r}
x - 5 \\
x - 1 \overline{)\, x^2 - 6x + 2} \\
x^2 - x \\
\end{array}
$$

$$
\begin{array}{r}
-5x + 2 \\
-5x + 5 \\
\hline
-3
\end{array}
$$

$P(1) = 1 - 6 + 2 = -3$

3.
$$
\begin{array}{r}
x^2 + x - 4 \\
x + 1 \overline{)\, x^3 + 2x^2 - 3x + 4} \\
x^3 + x^2 \\
\end{array}
$$

$$
\begin{array}{r}
x^2 - 3x \\
x^2 + x \\
\hline
-4x + 4 \\
-4x - 4 \\
\hline
8
\end{array}
$$

$P(-1) = (-1)^3 + 2(-1)^2 - 3(-1) + 4 = 8$

5.
$$
\begin{array}{r}
2x^2 - 8x + 17 \\
x + 2 \overline{)\, 2x^3 - 4x^2 + x - 1} \\
2x^3 + 4x^2 \\
\end{array}
$$

$$
\begin{array}{r}
-8x^2 + x \\
-8x^2 - 16x \\
\hline
17x - 1 \\
17x + 34 \\
\hline
-35
\end{array}
$$

$P(-2) = 2(-2)^3 - 4(-2)^2 + (-2) - 1 = -35$

7.

$$
\begin{array}{r}
3x^3 \;-\; 6x^2 \;+\; 8x \;-\; 16 \\
x+1{\overline{\smash{\big)}\,3x^4 \;-\; 3x^3 \;+\; 2x^2 \;-\; 8x \;+\; 1}} \\
\underline{3x^4 \;+\; 3x^3} \\
-6x^3 \;+\; 2x^2 \\
\underline{-6x^3 \;-\; 6x^2} \\
8x^2 \;-\; 8x \\
\underline{8x^2 \;+\; 8x} \\
-16x \;+\; 1 \\
\underline{-16x \;-\; 16} \\
17
\end{array}
$$

$$P(-1) = 3(-1)^3 - 6(-1)^2 + 8(-1) - 16 = 17$$

9.

$$
\begin{array}{r}
x^3 \;-\; 2x^2 \;+\; 2x \;-\; 1 \\
x+2{\overline{\smash{\big)}\,x^4 \qquad\quad -\; 2x^2 \;+\; 3x \;-\; 2}} \\
\underline{x^4 \;+\; 2x^3} \\
-2x^3 \;-\; 2x^2 \\
\underline{-2x^3 \;-\; 4x^2} \\
2x^2 \;+\; 3x \\
\underline{2x^2 \;+\; 4x} \\
-\; x \;-\; 2 \\
\underline{-\; x \;-\; 2} \\
0
\end{array}
$$

$$P(-2) = (-2)^4 - 2(-2)^2 + 3(-2) - 2 = 0$$

11.

$$
\begin{array}{r}
2x^2 \;-\; 3x \;+\; 2 \\
x-3{\overline{\smash{\big)}\,2x^3 \;-\; 9x^2 \;+\; 11x \;-\; 6}} \\
\underline{2x^3 \;-\; 6x^2} \\
-3x^2 \;+\; 11x \\
\underline{-3x^2 \;+\; 9x} \\
2x \;-\; 6 \\
\underline{2x \;-\; 6} \\
0
\end{array}
$$

$$P(3) = 2(3)^3 - 9(3)^2 + 11(3) - 6 = 0$$

13.

$$
\begin{array}{r}
-x^2 + 4x - 2 \\
x + 1 \overline{\smash{)}\; -x^3 + 3x^2 + 2x - 7} \\
\underline{-x^3 - x^2} \\
4x^2 + 2x \\
\underline{4x^2 + 4x} \\
-2x - 7 \\
\underline{-2x - 2} \\
-5
\end{array}
$$

$$P(-1) = -(-1)^3 + 3(-1)^2 + 2(-1) - 7 = -5$$

15.

$$
\begin{array}{r}
x^2 + 7x + 10 \\
3x - 2 \overline{\smash{)}\; 3x^3 + 19x^2 + 16x - 20} \\
\underline{3x^3 - 2x^2} \\
21x^2 + 16x \\
\underline{21x^2 - 14x} \\
30x - 20 \\
\underline{30x - 20}
\end{array}
$$

$$P\left(\tfrac{2}{3}\right) = 3\left(\tfrac{2}{3}\right)^3 + 19\left(\tfrac{2}{3}\right)^2 + 16\left(\tfrac{2}{3}\right) - 20 = 0$$

17.

$$
\begin{array}{r}
3x^2 - 10 \\
x^2 + 1 \overline{\smash{)}\; 3x^4 - 7x^2 + 1} \\
\underline{3x^4 + 3x^2} \\
-10x^2 + 1 \\
\underline{-10x^2 - 10} \\
11
\end{array}
$$

19.

$$
\begin{array}{r}
2x - 1 \\
x^2 - 3x + 3 \overline{\smash{)}\; 2x^3 - 7x^2 + 9x - 3} \\
\underline{2x^3 - 6x^2 + 6x} \\
-x^2 + 3x - 3 \\
\underline{-x^2 + 3x - 3}
\end{array}
$$

21. $P(4) = 4^2 - 5(4) + 4 = 0,$ Yes

23. $P(-1) = (-1)^3 - 1 = -2,$ No

25. $P(2) = 4(2)^3 + (2)^2 - 16(2) - 4 = 0,$ Yes

27. $P(2) = 3(2)^5 + 7(2) - 8 = 102,$ No

29. $P(4) = 4^4 + 2(4)^3 - 15(4)^2 - 32(4) - 16 = 0,$ Yes

31. $P\left(\frac{1}{3}\right) = 3\left(\frac{1}{3}\right)^3 + 2\left(\frac{1}{3}\right)^2 - 4\left(\frac{1}{3}\right) + 1 = 0,$ Yes

33. $P(-2i) = (-2i)^4 - 16 = 0,$ Yes

35. $P(2i) = (2i)^3 - 2(2i)^2 + 4(2i) - 8 = 0,$ Yes

37. $a(x-1)(x+1)(x-2) = P(x)$
$\quad a(-1)(1)(-2) = -2 = P(0)$
$\qquad\qquad a = -1$
$\quad -1(x-1)(x+1)(x-2) = P(x)$

39. $a(x-1)^2(x-2)^2 = P(x)$
$\quad P(0) = a(1)(4) = 3$
$\qquad\qquad a = \frac{3}{4}$
$\quad \frac{3}{4}(x-1)^2(x-2)^2 = P(x)$

41. $a(x-2+i)(x-2+i)(x-\sqrt{2}) = P(x)$
$\quad P(0) = a(-2+i)(-2-i)(-\sqrt{2}) = 2$
$\qquad\qquad -5\sqrt{2}a = 2, \quad a = -\frac{\sqrt{2}}{5}$
$\quad P(x) = -\frac{\sqrt{2}}{5}(x^2+5)(x-\sqrt{2})$

43. $\begin{vmatrix} -\lambda & -\frac{3}{5} & 0 \\ \frac{5}{3} & -\lambda & -\frac{5}{3} \\ 0 & 6 & -(\lambda+6) \end{vmatrix} = -\lambda \begin{vmatrix} -\lambda & -\frac{5}{3} \\ 6 & -(\lambda+6) \end{vmatrix} - \frac{5}{3} \begin{vmatrix} -\frac{3}{5} & 0 \\ 6 & -(\lambda+6) \end{vmatrix}$

$\qquad = -\lambda\left(\lambda(\lambda+6)+10\right) - \frac{5}{3}\left(\frac{3}{5}(\lambda+6)\right)$

$\qquad = -\lambda^3 - 6\lambda^2 - 10\lambda - \lambda - 6 = -\lambda^3 - 6\lambda^2 - 11\lambda - 6$

$\quad P(-1) = -(-1)^3 - 6(-1)^2 - 11(-1) - 6 = 0$

1.

$$4 \,\big)\, \begin{array}{ccc} 1 & -5 & 4 \\ & 4 & -4 \\ \hline 1 & -1 \end{array}$$

$$x - 1$$

3.

$$2 \,\big)\, \begin{array}{cccc} 1 & -2 & 3 & -1 \\ & 2 & 0 & 6 \\ \hline 1 & 0 & 3 & 5 \end{array}$$

$$x^2 + 3 + \frac{5}{x-2}$$

5.

$$-\tfrac{1}{2} \,\big)\, \begin{array}{cccc} 1 & 0 & -3 & 9 \\ & -\tfrac{1}{2} & \tfrac{1}{4} & \tfrac{11}{8} \\ \hline 1 & -\tfrac{1}{2} & -\tfrac{11}{4} & \tfrac{83}{8} \end{array}$$

$$x^2 - \frac{x}{2} - \frac{11}{4} + \frac{\frac{83}{8}}{x-\frac{1}{2}}$$

7.

$$3 \,\big)\, \begin{array}{ccccc} 1 & -1 & 2 & 0 & -72 \\ & 3 & 6 & 24 & 72 \\ \hline 1 & 2 & 8 & 24 \end{array}$$

$$x^3 + 2x^2 + 8x + 24$$

9.

$$2 \,\big)\, \begin{array}{cccccc} -3 & 0 & 0 & 0 & 7 & 8 \\ & -6 & -12 & -24 & -48 & -82 \\ \hline -3 & -6 & -12 & -24 & -41 & -74 \end{array}$$

$$-3x^4 - 6x^3 - 12x^2 - 24x - 41 - \frac{74}{x-2}$$

11.

$$-\tfrac{1}{3} \,\big)\, \begin{array}{cccc} 3 & -20 & 23 & 10 \\ & -1 & 7 & -10 \\ \hline 3 & -21 & 30 \end{array}$$

$$3x^2 - 21x + 30$$

13.

$$3 \,\big)\, \begin{array}{cccc} 1 & 0 & 2 & -5 \\ & 3 & 9 & 33 \\ \hline 1 & 3 & 11 & \underline{28} = P(3) \end{array}$$

$$P(3) = (3)^3 + 2(3) - 5 = 28$$

15.

$$\frac{1}{2} \overline{\left.\right)\ \begin{array}{cccc} 3 & 2 & -4 & 1 \end{array}}$$

$$\begin{array}{cccc} & \frac{3}{2} & \frac{7}{4} & -\frac{9}{8} \end{array}$$

$$\begin{array}{cccc} 3 & \frac{7}{2} & -\frac{9}{4} & \frac{1}{8} = P\left(\frac{1}{2}\right) \end{array}$$

$$P\left(\frac{1}{2}\right) = 3\left(\frac{1}{2}\right)^3 + 2\left(\frac{1}{2}\right)^2 - 4\left(\frac{1}{2}\right) + 1 = -\frac{1}{8}$$

17.

$$.1 \overline{\left.\right)\ \begin{array}{ccccc} -6 & 0 & 5 & 0 & -2 \end{array}}$$

$$\begin{array}{ccccc} -6 & -.6 & 4.94 & .494 & -1.9506 = P(.1) \end{array}$$

$$P(.1) = -6(.1)^4 + 5(.1)^2 - 2 = -1.9506$$

19.

$$3i \overline{\left.\right)\ \begin{array}{ccc} 1 & -2 & -1 \end{array}}$$

$$\begin{array}{ccc} & 3i & -6i-9 \end{array}$$

$$\begin{array}{ccc} 1 & -2+3i & -6i-10 = P(3i) \end{array}$$

$$P(3i) = (3i)^2 - 2(3i) - 1 = -6i - 10$$

21.

$$-2+i \overline{\left.\right)\ \begin{array}{cccc} 1 & 5 & 9 & 5 \end{array}}$$

$$\begin{array}{cccc} & -2+i & -7+i & -5 \end{array}$$

$$\begin{array}{cccc} 1 & 3+i & 2+i & 0 \ = P(-2+i) \end{array}$$

$$P(-2+i) = (-2+i)^3 + 5(-2+i)^2 + 9(-2+i) + 5$$

23. $P(1) = (1)^3 - 2(1)^2 + 3k(1) - 3$

$$= 1 - 2 + 3k - 3 = 0, \quad k = \frac{4}{3}$$

25.
$$P(\sqrt{2}) = (\sqrt{2})^3 - 2k(\sqrt{2}) + 5 = 0$$
$$-2\sqrt{2}k = -5 - 2\sqrt{2}$$
$$k = \frac{5 + 2\sqrt{2}}{2\sqrt{2}} = 1 + \frac{5}{2\sqrt{2}}$$

27. $P(x) = x^3 + 8x + 12$

$s_1 = 1(2) = 2$

$s_2 = 2 + 0 = 2$

$s_3 = 2(2) = 4$

$s_4 = 4 + 8 = 12$

$s_5 = 12(2) = 24$

$s_6 = 24 + 12 = 36$

29. $P(x) = 3x^7 - 2x^4 + 7x^2 - 5$

$s_1 = 3(-3) = -9$ $s_8 = 249 + 0$

$s_2 = -9 + 0 = -9$ $s_9 = 249(-3) = -747$

$s_3 = -9(-3) = 27$ $s_{10} = -747 + 7 = -740$

$s_4 = 27 + 0 = 27$ $s_{11} = -740(-3) = 2220$

$s_5 = 27(-3) = -81$ $s_{12} = 2220 + 0$

$s_6 = -81 + (-2) = -83$ $s_{13} = 2220(-3) = -6660$

$s_7 = (-83)(-3) = 249$ $s_{14} = -6660 - 5 = -6665 = P(-3)$

31. $s_1 = 2(2.4) = 4.8$

$s_2 = 4.8 + (-3) = 1.8$

$s_3 = (1.8)(2.4) = 4.32$

$s_4 = 4.32 + 0 = 4.32$

$s_5 = 4.32(2.4) = 10.368$

$s_6 = 10.368 + 5 = 15.368$

$s_7 = 15.368(2.4) = 36.8832$

$s_8 = 36.8832 + 25 = 61.8832$

1. $x^3 + 2x^2 - x - 2 = 0$

   Possible ratio. roots: $\pm 1, 2$

   $$
   \begin{array}{r|rrrr}
   1 & 1 & 2 & -1 & -2 \\
     &   & 1 & 3 & 2 \\
   \hline
     & 1 & 3 & 2 & 0
   \end{array}
   $$

   $(x-1)(x^2 + 3x + 2) = 0$

   $(x-1)(x+1)(x+2) = 0$

   $x = 1, -1, -2$

3. $y^3 + 2y^2 + y + 2 = 0$

   Possible ratio. roots: $\pm 1, 2$

   $$
   \begin{array}{r|rrrr}
   -2 & 1 & 2 & 1 & 2 \\
      &   & -2 & 0 & -2 \\
   \hline
      & 1 & 0 & 1 &
   \end{array}
   $$

   $(y+2)(y^2 + 1) = 0$

   $y = -2, \pm i$

5. $2x^3 + x^2 - 18x - 20 = 0$

   Possible ratio. roots: $\pm 1, 2, 4, 5, 10, \frac{1}{2}, \frac{5}{2}$

   $$
   \begin{array}{r|rrrr}
   -\frac{5}{2} & 2 & 1 & -18 & 20 \\
                &   & -5 & 10 & 20 \\
   \hline
                & 2 & -4 & -8 &
   \end{array}
   $$

   $(2x^2 - 4x - 8)\left(x+\frac{5}{2}\right) = 0$

   $x = -\frac{5}{2}, \quad \dfrac{4 \pm \sqrt{16+64}}{4}$

   $x = -\frac{5}{2}, \quad 1 \pm \sqrt{5}$

7. $z^4 - 2z^3 - 7z^2 + 8z + 12 = 0$

Possible ratio. roots: $\pm 1, 2, 3, 4, 6, 12$

$$
\begin{array}{r|rrrrr}
-1 \,) & 1 & -2 & -7 & 8 & 12 \\
& & -1 & 3 & 4 & -12 \\
\hline
& 1 & -3 & -4 & 12 & 0
\end{array}
$$

$(x+1)(x^3 - 3x^2 - 4x + 12) = 0$

$$
\begin{array}{r|rrrr}
2 \,) & 1 & -3 & -4 & 12 \\
& & 2 & -2 & -12 \\
\hline
& 1 & -1 & -6 & 0
\end{array}
$$

$$(x+1)(x-2)(x^2-x-6) = 0$$
$$(x+1)(x-2)(x-3)(x+2) = 0$$
$$x = -1, -2, 2, 3$$

9. $3p^3 - 10p^2 + 12p - 3 = 0$

Possible ratio. roots: $\pm 1, 3, \frac{1}{3}$

$$
\begin{array}{r|rrrr}
\frac{1}{3} \,) & 3 & -10 & 12 & -3 \\
& & 1 & -3 & 3 \\
\hline
& 3 & -9 & 9 & 0
\end{array}
$$

$$\left(x - \frac{1}{3}\right)3(x^2 - 3x + 3) = 0$$
$$x = \frac{1}{3}, \quad \frac{3 \pm \sqrt{9-12}}{2}$$
$$x = \frac{1}{3}, \quad \frac{3 \pm i\sqrt{3}}{2}$$

11. $x^4 - 6x^2 - 8x - 3 = 0$

Possible ratio. roots: $\pm 1, 3$

$$
\begin{array}{r|rrrrr}
-1 & 1 & 0 & -6 & -8 & -3 \\
   &   & -1 & 1 & 5 & 3 \\
\hline
   & 1 & -1 & -5 & -3 & 0 \\
\end{array}
$$

$$(x+1)(x^3 - x^2 - 5x - 3) = 0$$

$$
\begin{array}{r|rrrr}
-1 & 1 & -1 & -5 & -3 \\
   &   & -1 & 2 & 3 \\
\hline
   & 1 & -2 & -3 & 0 \\
\end{array}
$$

$$(x+1)^2(x^2 - 2x - 3) = 0$$
$$(x+1)^2(x+1)(x-3) = 0$$
$$(x+1)^3(x-3) = 0$$
$$x = -1, -1, -1, 3$$

13. $9w^3 - w + 2 = 0$

Possible ratio. roots: $\pm 1, 2, \frac{1}{9}, \frac{1}{3}, \frac{2}{9}, \frac{2}{3}$

$$
\begin{array}{r|rrrr}
-\frac{2}{3} & 9 & 0 & -1 & 2 \\
   &   & -6 & 4 & -2 \\
\hline
   & 9 & -6 & 3 & \\
\end{array}
$$

$$\left(x+\frac{2}{3}\right)3(3x^2 - 2x + 1) = 0$$

$$x = -\frac{2}{3}, \quad \frac{2 \pm \sqrt{4-12}}{6}$$

$$x = -\frac{2}{3}, \quad \frac{1 \pm i\sqrt{2}}{3}$$

15. $2x^3 - 3x^2 - 5x - 2 = 0$

   Possible ratio. roots: $\pm 1, 2, \frac{1}{2}$

$$\frac{1}{2}\overline{\smash{\big)}\ \begin{array}{rrrr} 2 & -3 & 5 & -2 \\ & 1 & -1 & 2 \end{array}}$$
$$\begin{array}{rrrr} 2 & -2 & 4 & 0 \end{array}$$

$$\left(x - \tfrac{1}{2}\right)(x^2 - x + 2) = 0$$

$$x = \tfrac{1}{2}, \quad \frac{1 \pm \sqrt{1-8}}{2}$$

$$x = \tfrac{1}{2}, \quad \frac{1 \pm i\sqrt{7}}{2}$$

17. $x^5 - x^3 - 8x^2 + 8 = 0$

   Possible ratio. roots: $\pm 1, 2, 4, 8$

$$1\ \overline{\smash{\big)}\ \begin{array}{rrrrrr} 1 & 0 & -1 & -8 & 0 & 8 \\ & 1 & 1 & 0 & -8 & -8 \end{array}}$$
$$\begin{array}{rrrrrr} 1 & 1 & 0 & -8 & -8 \end{array}$$

$$(x-1)(x^4 + x^3 - 8x - 8) = 0$$

$$-1\ \overline{\smash{\big)}\ \begin{array}{rrrrr} 1 & 1 & 0 & -8 & -8 \\ & -1 & 0 & 0 & 8 \end{array}}$$
$$\begin{array}{rrrr} 1 & 0 & 0 & -8 \end{array}$$

$$(x-1)(x+1)(x^3-8) = 0$$

$$(x-1)(x+1)(x-2)(x^2+2x+4) = 0$$

$$x = 1, -1, 2, \quad \frac{-2 \pm \sqrt{4-16}}{2}$$

$$x = 1, -1, 2, -1 \pm i\sqrt{3}$$

19. $x^3 + 6x^2 + 2x - 3 = 0$

Possible ratio. roots: $\pm 1, 3$

$$-1 \overline{\big)\ \begin{array}{cccc} 1 & 6 & 2 & -3 \\ & -1 & -5 & 3 \end{array}}$$
$$\begin{array}{cccc} 1 & 5 & -3 \end{array}$$

$(x+1)(x^2 + 5x - 3) = 0$

$$x = -1, \quad \frac{-5 \pm \sqrt{25+12}}{2}$$

$$x = -1, \quad \frac{-5 \pm \sqrt{37}}{2}$$

$$(x+1)\left(x+\frac{5}{2}+\frac{\sqrt{37}}{2}\right)\left(x+\frac{5}{2}-\frac{\sqrt{37}}{2}\right)$$

21. $4x^3 + 8x^2 + 9x + 3 = 0$

Possible ratio. roots: $\pm 1, 3, \frac{1}{4}, \frac{3}{4}, \frac{1}{2}$

$$-\frac{1}{2} \overline{\big)\ \begin{array}{cccc} 4 & 8 & 9 & 3 \\ & -2 & -3 & -3 \end{array}}$$
$$\begin{array}{ccc} 4 & 6 & 6 \end{array}$$

$\left(x-\frac{1}{2}\right)2(2x^2 + 3x + 3) = 0$

$$x = -\frac{1}{2}, \quad \frac{-3 \pm i\sqrt{15}}{4}$$

$$2(2x+1)\left(x+\frac{3}{4}+\frac{i\sqrt{15}}{4}\right)\left(x+\frac{3}{4}-\frac{i\sqrt{15}}{4}\right)$$

23. $4x^4 - 4x^3 - 57x^2 + 81x = 0$

$x(4x^3 - 4x^2 - 57x + 81) = 0$

Possible ratio. roots: $\pm 1, 3, 9, 27, 81, \frac{1}{2}, \frac{3}{2}, \frac{9}{2}, \frac{27}{2}, \frac{81}{2}, \frac{1}{4}, \frac{3}{4}, \frac{9}{4}, \frac{27}{4}, \frac{81}{4}$

$$\frac{3}{2} \overline{\big)\ \begin{array}{cccc} 4 & -4 & -57 & 81 \\ & 6 & 3 & -81 \end{array}}$$
$$\begin{array}{ccc} 4 & 2 & -54 \end{array}$$

$x\left(x-\frac{3}{2}\right)2(2x^2 + x - 27) = 0$

$$x = 0, \frac{3}{2}, \quad \frac{-1 \pm \sqrt{217}}{4}$$

$$2x(2x-3)\left(x+\frac{1}{4}+\frac{\sqrt{217}}{4}\right)\left(x+\frac{1}{4}-\frac{\sqrt{217}}{4}\right)$$

25. $y = x^4$

$y = 2x - x^3$

$$x^4 = 2x - x^3$$

$$x^4 + x^3 - 2x = 0$$

$$x(x^3 + x^2 - 2) = 0$$

Possible ratio. roots: $\pm 1, 2$

$$
\begin{array}{r|rrrr}
1 & 1 & 1 & 0 & -2 \\
  &   & 1 & 2 & 2 \\
\hline
  & 1 & 2 & 2 &
\end{array}
$$

$$x(x-1)(x^2 + 2x + 2) = 0$$

$$x = 0, \quad 1, \quad -1 \pm i$$

$x = 0, \quad y = 0$ $x = -1 + i, \quad y = (-1 + i)^4 = -4$

$x = 1, \quad y = 1$ $x = -1 - i, \quad y = (-1 - i)^4 = -4$

27. $x^2 - xy = y + 1$

$xy + x = 4y^2$

$x^2 - 1 = y + xy$

$x^2 - 1 = y(1+x)$

$\dfrac{x^2-1}{1+x} = y$

$y = x-1, \quad x \neq -1$

$xy + x = 4y^2$

$x(x-1) + x = 4(x-1)^2$

$x^2 - x + x = 4x^2 - 8x + 4$

$3x^2 - 8x + 4 = 0$

$(3x-2)(x-2) = 0$

$x = 2, \quad \frac{2}{3}$

$x = 2, \quad y = 2-1 = 1$ $(2, 1)$

$x = \frac{2}{3}, \quad y = \frac{2}{3}-1 = -\frac{1}{3}$ $\left(\frac{2}{3}, -\frac{1}{3}\right)$

Consider $x = -1$

$x = -1, \quad -y - 1 = 4y^2$

$4y^2 + y + 1 = 0$

$$y = \frac{-1 \pm \sqrt{15i}}{8}$$

1. $x^2 + 3x + 2 = 0$ no positive

   $P(-x) = -x^3 - 3x + 2$ 1 negative, 2 complex

3. $x^4 + 13x^2 - 7 = 0$ 1 positive

   $P(-x) = x^4 + 13x^2 - 7$ 1 negative

                        So 2 complex

5. $2x^3 - 7x^2 - 3x - 1 = 0$ 1 positive

   $P(-x) = -2x^3 - 7x^2 + 3x - 1$ 2 or 0 negative

   So 1 positive, 2 complex   or   1 positive, 2 negative

7. $-2x^5 + 8x^3 - 7x + 15 = 0$ 3 or 1 positive

   $P(-x) = 2x^5 - 8x^3 + 7x + 15$ 2 or 0 negative

   3 positive, 2 complex   or   3 positive, 2 negative   or
   1 positive, 4 complex   or   1 positive, 2 negative, 2 complex

9. $3x^6 + x^4 - 8x^2 + 16 = 0$ 2 or 0 positive

   $P(-x) = 3x^6 + x^4 - 8x^2 + 16$ 2 or 0 negative

   2 positive, 4 complex   or   2 positive, 2 negative, 2 complex   or
   2 negative, 4 complex   or   6 complex

11. $x^3 + 2x^2 + 8x - 3 = 0$

    $P(0) = -3, \quad P(1) = 8$

    $c = \dfrac{0(8) - 1(-3)}{8 + 3} = \dfrac{3}{11} = .27$

13. $x^3 - 3x + 1 = 0$

    $P(1) = -1, \quad P(2) = 3$

    $c = \dfrac{1(3) - 2(-1)}{3 + 1} = \dfrac{5}{4} = 1.25$

15. $2x^3 - 3x^2 - 12x + 6 = 0$

$P(3) = 2(27) - 3(9) - 12(3) + 6 = -3$

$P(4) = 2(64) - 3(16) - 12(4) + 6 = 38$

$c = \dfrac{3(38) - 4(-3)}{38 + 3} = \dfrac{126}{41} = 3.07$

17. $x^4 - x^3 - 5x^2 + 3x + 2 = 0$

$P(-1) = 1 + 1 - 5 - 3 + 2 = -4$

$P(0) = 2$

$c = \dfrac{-1(2) - 0(-4)}{2 + 4} = -\dfrac{2}{6} = -.33$

19. $x^4 - x^3 - 2x^2 - x - 3 = 0$

$P(2) = 16 - 8 - 8 - 2 - 3 = -5$

$P(3) = 81 - 27 - 2(9) - 3 - 3 = 24$

$c = \dfrac{2(24) - 3(-5)}{24 + 5} = \dfrac{63}{29} = 2.17$

21. $2.1x^3 + 3.4x^2 - 1.8x + 5.3 = 0$

$P(-3) = 2.1(-27) + 3.4(9) - 1.8(-3) + 5.3 = -15.4$

$P(-2) = 2.1(-8) + 3.4(-4) - 1.8(-2) + 5.3 = 15.7$

$c = \dfrac{-3(5.7) + 2(-15.4)}{5.7 + 15.4} = -\dfrac{47.9}{21.1} = -2.27$

Section 12.6   Decomposition of Rational Functions into Partial Fractions

1. $\dfrac{x + 1}{(x+5)(x-1)} = \dfrac{A}{x + 5} + \dfrac{B}{x - 1}$

$x + 1 = A(x-1) + B(x+5)$

$x + 1 = Ax - A + Bx + 5B$

$x + 1 = (A+B)x + (-A + 5B)$

$\begin{aligned} A + B &= 1 \\ -A + 5B &= 1 \end{aligned}$

$6B = 2, \quad B = \tfrac{1}{3}, \quad A = \tfrac{2}{3}$

3. $\dfrac{1}{x(x+1)} = \dfrac{A}{x} + \dfrac{B}{x+1}$

$1 = A(x+1) + Bx$

$1 = (A+B)x + A$

$A = 1, \quad A + B = 0, \quad B = -1$

5. $\dfrac{2x^2 + 5x + 5}{x(x+5)(x+2)} = \dfrac{A}{x} + \dfrac{B}{x+5} + \dfrac{C}{x+2}$

$2x^2 + 5x + 5 = A(x+5)(x+2) + Bx(x+2) + Cx(x+5)$

$2x^2 + 5x + 5 = Ax^2 + 7Ax + 10A + Bx^2 + 2Bx + Cx^2 + 5Cx$

$2x^2 + 5x + 5 = (A+B+C)x^2 + (7A+2B+5C)x + 10A$

$A + B + C = 2$

$7A + 2B + 5C = 5$

$10A = 5, \quad A = \dfrac{1}{2}$

$B + \quad C = \dfrac{3}{2}$

$\underline{2B + 5C = \dfrac{3}{2}}$

$-2B - 2C = -3$

$\underline{2B + 5C = \dfrac{3}{2}}$

$3C = -\dfrac{3}{2}, \quad C = -\dfrac{1}{2}, \quad B = 2$

7. $\dfrac{x+3}{x^2 + 6x + 5} = \dfrac{x+3}{(x+5)(x+1)} = \dfrac{A}{x+5} + \dfrac{B}{x+1}$

$x + 3 = A(x+1) + B(x+5)$

$x + 3 = Ax + A + Bx + 5B$

$x + 3 = (A+B)x + A + 5B$

$A + \quad B = \quad 1$

$\underline{A + 5B = \quad 3}$

$-4B = -2, \quad B = \dfrac{1}{2}, \quad A = \dfrac{1}{2}$

9. $\dfrac{x^2 + 3x - 6}{x(x-1)^2} = \dfrac{A}{x} + \dfrac{B}{x-1} + \dfrac{C}{(x-1)^2}$

$x^2 + 3x - 6 = A(x-1)^2 + Bx(x-1) + Cx$

$x^2 + 3x - 6 = Ax^2 - 2Ax + A + Bx^2 - Bx + Cx$

$x^2 + 3x - 6 = (A+B)x^2 + (-2A - B + C)x + A$

$A + B = 1$

$-2A - B + C = 3$

$A = -6, \quad B = 7$

$C = 3 + 2A + B = -2$

11. $\dfrac{1}{x(x^2 + 2x + 1)} = \dfrac{1}{x(x+1)^2} = \dfrac{A}{x} + \dfrac{B}{x+1} + \dfrac{C}{(x+1)^2}$

$1 = A(x+1)^2 + Bx(x+1) + Cx$

$1 = Ax^2 + 2Ax + A + Bx^2 + Bx + Cx$

$1 = (A+B)x^2 + (2A+B+C) + A$

$A + B = 0$

$2A + B + C = 0$

$A = 1, \quad B = -1$

$C = -2A - B = -1$

13. $\dfrac{x^2 - 3x + 6}{(x-1)(x+1)(3-2x)} = \dfrac{A}{x-1} + \dfrac{B}{x+1} + \dfrac{C}{3-2x}$

$x^2 - 3x + 6 = A(x+1)(3-2x) + B(x-1)(3-2x) + C(x-1)(x+1)$

$x^2 - 3x + 6 = -2Ax^2 + Ax + 3A - 2Bx^2 + 5Bx - 3B + Cx^2 - C$

$x^2 - 3x + 6 = (-2A-2B+C)x^2 + (A+5B)x + (3A-3B-C)$

$\begin{array}{r} -2A - 2B + C = 1 \\ A + 5B = -3 \\ \underline{3A - 3B - C = 6} \end{array}$

$\begin{array}{r} -2A - 2B + C = 1 \\ \underline{3A - 3B - C = 6} \\ A - 5B = 7 \\ \underline{A + 5B = -3} \end{array}$

$2A = 4, \quad A = 2$

$B = \dfrac{-3 - A}{5} = -\dfrac{5}{5} = -1, \quad C = 1 + 2A + 2B = 3$

15. $\dfrac{2x-1}{(x-1)^2} = \dfrac{A}{x-1} + \dfrac{B}{(x+1)^2}$

$2x+1 = A(x-1) + B$

$2x-1 = Ax - A + B$

$A = 2, \quad -A + B = -1$

$\qquad\qquad B = -1 + A = 1$

17. $\dfrac{x^2 + x + 2}{x(x^2+2)} = \dfrac{A}{x} + \dfrac{Bx+C}{x^2+2}$

$x^2 + x + 2 = A(x^2+2) + (Bx+C)x$

$x^2 + x + 2 = Ax^2 + 2A + Bx^2 + Cx$

$x^2 + x + 2 = (A+B)x^2 + Cx + 2A$

$\qquad A + B = 1$

$\qquad\qquad C = 1$

$\qquad 2A = 2, \quad A = 1, \quad B = 0$

19. $\dfrac{x^2+1}{(2x+3)(x^2+2x+4)} = \dfrac{A}{2x+3} + \dfrac{Bx+C}{x^2+2x+4}$

$x^2 + 1 = A(x^2+2x+4) + (Bx+C)(2x+3)$

$x^2 + 1 = Ax^2 + 2Ax + 4A + 2Bx^2 + 2Cx + 3Bx + 3C$

$x^2 + 1 = (A+2B)x^2 + (2A+3B+2C)x + (4A+3C)$

$\qquad\qquad A + 2B = 1$

$\qquad 2A + 3B + 2C = 0$

$\qquad\qquad 4A + 3C = 1$

$$A = \dfrac{\begin{vmatrix} 1 & 2 & 0 \\ 0 & 3 & 2 \\ 1 & 0 & 3 \end{vmatrix}}{\begin{vmatrix} 1 & 2 & 0 \\ 2 & 3 & 2 \\ 4 & 0 & 3 \end{vmatrix}} = \dfrac{13}{13} = 1$$

$$B = \dfrac{\begin{vmatrix} 1 & 1 & 0 \\ 2 & 0 & 2 \\ 4 & 1 & 3 \end{vmatrix}}{13} = \dfrac{0}{13} = 0$$

$$C = \dfrac{\begin{vmatrix} 1 & 2 & 1 \\ 2 & 3 & 0 \\ 4 & 0 & 1 \end{vmatrix}}{13} = -\dfrac{13}{13} = -1$$

21. $\dfrac{x^3 - 3x^2 + 7x - 5}{(x^2 + 2x + 3)(x^2 + 5)} = \dfrac{Ax + B}{x^2 + 2x + 3} + \dfrac{Cx + D}{x^2 + 5}$

$x^3 - 3x^2 + 7x - 5 = (Ax+B)(x^2+5) + (Cx+D)(x^2+2x+3)$

$x^3 - 3x^2 + 7x - 5 = Ax^3 + 5Ax + Bx^2 + 5B + Cx^3 + 2Cx^2 + 3Cx + Dx^2 + 2Dx + 3D$

$x^3 - 3x^2 + 7x - 5 = (A+C)x^3 + (B+2C+D)x^2 + (5A+3C+2D)x + (5B+3D)$

$$A + C = 1$$
$$B + 2C + D = -3$$
$$5A + 3C + 2D = 7$$
$$5A + 3D = -5$$

$$A = \dfrac{\begin{vmatrix} 1 & 0 & 1 & 0 \\ -3 & 1 & 2 & 1 \\ 7 & 0 & 3 & 2 \\ -5 & 5 & 0 & 3 \end{vmatrix}}{\begin{vmatrix} 1 & 0 & 1 & 0 \\ 0 & 1 & 2 & 1 \\ 5 & 0 & 3 & 2 \\ 0 & 5 & 0 & 3 \end{vmatrix}} = \dfrac{1\begin{vmatrix} 1 & 2 & 1 \\ 0 & 3 & 2 \\ 5 & 0 & 3 \end{vmatrix} + 1\begin{vmatrix} -3 & 1 & 1 \\ 7 & 0 & 2 \\ -5 & 5 & 3 \end{vmatrix}}{1\begin{vmatrix} 1 & 2 & 1 \\ 0 & 3 & 2 \\ 5 & 0 & 3 \end{vmatrix} + 1\begin{vmatrix} 0 & 1 & 1 \\ 5 & 0 & 2 \\ 0 & 5 & 3 \end{vmatrix}} = \dfrac{48}{24} = 2$$

$$B = \dfrac{\begin{vmatrix} 1 & 1 & 1 & 0 \\ 0 & -3 & 2 & 1 \\ 5 & 7 & 3 & 2 \\ 0 & -5 & 0 & 3 \end{vmatrix}}{24} = \dfrac{-24}{24} = -1$$

$$C = \dfrac{\begin{vmatrix} 1 & 0 & 1 & 0 \\ 0 & 1 & -3 & 1 \\ 5 & 0 & 7 & 2 \\ 0 & 5 & -5 & 3 \end{vmatrix}}{24} = \dfrac{-24}{24} = -1$$

$$D = \dfrac{\begin{vmatrix} 1 & 0 & 1 & 1 \\ 0 & 1 & 2 & -3 \\ 5 & 0 & 3 & 7 \\ 0 & 5 & 0 & -5 \end{vmatrix}}{24} = \dfrac{0}{24} = 0$$

1.

$$\begin{array}{r|rrr} 1 & 1 & -3 & 4 \\ & & 1 & -2 \\ \hline & 1 & -2 & 2 \end{array}$$

$$x^2 - 3x + 4 \div (x-1) = x - 2 + \frac{2}{x-1}$$

$$P(1) = 1 - 3 + 4 = 2$$

3.

$$\begin{array}{r|rrr} 2 & 3 & -1 & 4 \\ & & 6 & 10 \\ \hline & 3 & 5 & 14 \end{array}$$

$$3x^2 - x + 4 \div (x-2) = 3x + 5, \quad R = 14$$

$$P(2) = 3(4) - 2 + 4 = 14$$

5.

$$\begin{array}{r|rrrr} \frac{1}{2} & 2 & 0 & -1 & 1 \\ & & 1 & \frac{1}{2} & \frac{1}{4} \\ \hline & 2 & 1 & -\frac{1}{2} & \frac{3}{4} \end{array}$$

$$2x^3 - x + 1 \div (2x-1) = 2x^2 + x - \frac{1}{2}, \quad R = \frac{3}{4}$$

$$P\left(\frac{1}{2}\right) = 2\left(\frac{1}{8}\right) - \frac{1}{2} + 1 = \frac{3}{4}$$

7.

$$\begin{array}{r|rrrrrr} -1 & 3 & 1 & 0 & 1 & -1 & -2 \\ & & -3 & 2 & -2 & 1 & 0 \\ \hline & 3 & -2 & 2 & -1 & 0 & -2 \end{array}$$

$$3x^5 + x^4 + x^2 - x - 2 \div (x+1) = 3x^4 - 2x^3 + 2x^2 - x, \quad R = -2$$

$$P(-1) = -3 + 1 + 1 - 2 = -2$$

9.

$$\begin{array}{r} \frac{2}{3}x^3 + \frac{2}{9}x^2 + \frac{2}{27}x + \frac{29}{81} \end{array}$$

$$3x - 1 \overline{\smash{\big)}\ 2x^4 \qquad\qquad\qquad + \quad x \quad - \quad 1}$$

$$\underline{2x^4 - \frac{2x^3}{3}}$$

$$\frac{2}{3}x^3$$

$$\underline{\frac{2}{3}x^3 - \frac{2x^2}{9}}$$

$$\frac{2x^2}{9} + x$$

$$\underline{\frac{2x^2}{9} - \frac{2}{27}x}$$

$$\frac{29}{27}x - 1$$

$$\underline{\frac{29}{27}x - \frac{29}{81}}$$

$$-\frac{52}{81} = R$$

$$P\left(\tfrac{1}{3}\right) = 2\left(\tfrac{1}{81}\right) + \tfrac{1}{3} - 1 = -\tfrac{52}{81}$$

11. $P(3) = 3^2 - 3(3) + 2 = +2,$ No factor

13. $P(2) = 2^4 - 16 = 0,$ Yes

15. $P\left(-\tfrac{1}{2}\right) = 2\left(-\tfrac{1}{2}\right)^3 - \left(-\tfrac{1}{2}\right)^2 - \tfrac{1}{2} - 1 = -2,$ No factor

17. $P(x) = a(x-2)(x-2-i)(x-2+i)(x-i)(x+i)$

$P(0) = 1 = a(-2)(-2-i)(-2+i)(-i)(i),$ $a = -\tfrac{1}{10}$

$P(x) = -\tfrac{1}{10}(x-2)(x^2-4x+5)(x^2+1)$

19. $P(x) = a(x-i)(x+i)(x-2i)(x+2i)$

$P(1) = 20 = a(1-i)(1+i)(1-2i)(1+2i),$ $a = 2$

$P(x) = 2(x^2+1)(x^2+4)$

21. $P(x) = a(x+2)(x+1)(x-1)(x-3)(x-4)$

$P(0) = -25 = a(2)(1)(-1)(-3)(-4),$ $a = \tfrac{25}{24}$

$P(x) = \tfrac{25}{24}(x+2)(x+1)(x-1)(x-3)(x-4)$

23. $x^4 + x^3 - x^2 + x - 2 = 0$

Possible ratio. roots: $\pm 1, 2$

$$
\begin{array}{r|rrrrr}
1 & 1 & 1 & -1 & 1 & -2 \\
  &   & 1 & 2 & 1 & 2 \\
\hline
\end{array}
$$

$$
\begin{array}{r|rrrr}
-2 & 1 & 2 & 1 & 2 \\
   &   & -2 & 0 & -2 \\
\hline
   & 1 & 0 & 1 \\
\end{array}
$$

$(x-1)(x+2)(x^2+1) = 0$

$x = 1, -2, \pm i$

25. $x^4 + x^3 - 3x^2 - x + 2 = 0$

Possible ratio. roots: $\pm 1, 2$

$$
\begin{array}{r|rrrrr}
1 & 1 & 1 & -3 & -1 & 2 \\
  &   & 1 & 2 & -1 & -2 \\
\hline
\end{array}
$$

$$
\begin{array}{r|rrrr}
1 & 1 & 2 & -1 & -2 \\
  &   & 1 & 3 & 2 \\
\hline
  & 1 & 3 & 2 \\
\end{array}
$$

$(x-1)^2(x^2+3x+2) = 0$

$(x-1)^2(x+2)(x+1) = 0$

$x = +1, +1, -2, -1$

27. $x^3 - 7x^2 + 5x + 1 = 0$

Possible ratio. roots: ± 1

$$
\begin{array}{r|rrrr}
1 & 1 & -7 & 5 & 1 \\
  &   & 1 & -6 & -1 \\
\hline
  & 1 & -6 & 1 \\
\end{array}
$$

$(x-1)(x^2-6x-1) = 0$

$x = 1, \quad \dfrac{6 \pm \sqrt{40}}{2} = 3 \pm \sqrt{10}$

29. $2x^3 - 3x^2 + 2x + 2 = 0$

Possible ratio. roots: $\pm 1, 2, \frac{1}{2}$

$$-\tfrac{1}{2} \;\overline{\big)\; \begin{array}{cccc} 2 & -3 & 2 & 2 \\ & -1 & 2 & 0 \end{array}}$$
$$\begin{array}{cccc} 2 & -4 & 4 \end{array}$$

$\left(x+\tfrac{1}{2}\right)2(x^2 - 2x + 2) = 0$

$x = -\tfrac{1}{2}, \quad \dfrac{2 \pm \sqrt{-4}}{2} = -\tfrac{1}{2}, \quad 1 \pm i$

31. $P(x) = x^4 - x - 3, \quad$ 1 positive

$P(-x) = x^4 + x - 3,$ 1 negative,

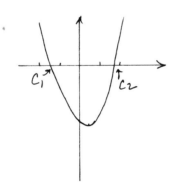

$P(2) = \;\;11$
$P(1) = -\;3$ $\left.\begin{array}{c}\\[2pt]\\\end{array}\right]$

$c_2 = \dfrac{1(11) - (2)(-3)}{11 + 3} = 1.21$

$P(0) = -3$

$P(-1) = -\;1$
$P(-2) = \;\;15$ $\left.\begin{array}{c}\\[2pt]\\\end{array}\right]$

$c_1 = \dfrac{-2(-1) - (-1)(15)}{-1 - 15} = -1.06$

33. $P(x) = x^5 - 1.1$,   1 positive

     $P(-x) = -x^5 - 1.1$,   0 negative

$P(0) = -1.1$

$P(1) = -.1$

$P(2) = 30.9$

$$c = \frac{1(30.9) - 2(-.1)}{30.9 + .1} = 1.003$$

35. $P(x) = 2.3x^3 - 7x + 1.8$,   2 positive

     $P(-x) = -2.3x^3 + 7x + 1.8$,   1 negative

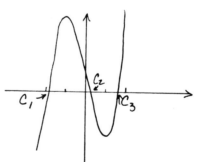

$P(-2) = -2.6$

$P(-1) = 6.5$

$$c_1 = \frac{-2(6.5) - (-1)(-2.6)}{6.5 + 2.6} = -1.71$$

$P(0) = 1.8$

$P(1) = -2.9$

$$c_2 = \frac{0(-2.9) - 1(1.8)}{-2.9 - 1.8} = .38$$

$P(1) = -2.9$

$P(2) = 6.2$

$$c_3 = \frac{1(6.2) - 2(2.9)}{6.2 + 2.9} = 1.32$$

37. $\dfrac{8x + 7}{(x-1)(x^2+4)} = \dfrac{A}{x - 1} + \dfrac{Bx + C}{x^2 + 4}$

$8x + 7 = A(x^2+4) + (Bx+C)(x-1)$

$8x + 7 = Ax^2 + 4A + Bx^2 - Bx + Cx - C$

$8x + 7 = (A+B)x^2 + (-B+C)x + (4A-C)$

$A + B = 0, \quad A = -B$

$-B + C = 8$

$4A - C = 7$

$\quad\quad -B + C = 8$

$\quad\quad \underline{-4B - C = 7}$

$\quad\quad -5B \quad\quad = 15, \quad B = -3, \quad A = 3, \quad C = 8+B = 5$

39. $\dfrac{x^2 + 3x - 9}{x^3 - x^2 + 4x - 4} = \dfrac{x^2 + 3x - 9}{x^2(x-1) + 4(x-1)} = \dfrac{x^2 + 3x - 9}{(x^2+4)(x-1)} = \dfrac{A}{x - 1} + \dfrac{Bx + C}{x^2 + 4}$

$x^2 + 3x - 9 = A(x^2+4) + (Bx+C)(x-1)$

$x^2 + 3x - 9 = Ax^2 + 4A + Bx^2 - Bx + Cx - C$

$x^2 + 3x - 9 = (A+B)x^2 + (C-B)x + (4A-C)$

$A + B = 1$

$C - B = 3$

$4A - C = -9$

$\quad\quad A + B = 1$

$\quad\quad \underline{C - B = 3}$

$\quad\quad A + C = 4$

$\quad\quad \underline{4A - C = -9}$

$\quad\quad 5A \quad\quad = -5, \quad A = -1, \quad B = 2, \quad C = 5$

41. $P(x) = x - \frac{1}{3}x^3$

$P(.1) = .1 - \frac{1}{3}(.1)^3 = .099667, \quad \sin .1 - .099833$

$P(.2) = .2 - \frac{1}{3}(.2)^3 = .197333, \quad \sin .2 = .198669$

$P(.5) = .5 - \frac{1}{3}(.5)^3 = .458333, \quad \sin .5 = .479426$

43. $\frac{1}{2} = \frac{1}{R_2 - 1} + \frac{1}{R_2} + \frac{1}{R_2 + 1}$

$(R_2-1)(R_2)(R_2+1) = 2R_2(R_2+1) + (R_2{}^2-1)2 + 2R_2(R_2-1)$

$R_2{}^3 - R_2 = 2R_2{}^2 + \cancel{2R_2} + 2R_2{}^2 - 2 + 2R_2{}^2 - \cancel{2R_2}$

$R_2{}^3 - 6R_2{}^2 - R_2 + 2 = 0$ 2 positive roots

$-R_2{}^3 - 6R_2{}^2 + R_2 + 2 = 0$ 1 negative root

Because $R_1 = R_2-1$ and $R_1 > 0$, $R_2 > 1$

$P(6) = -4$

$P(7) = 42$

$c = \frac{6(42) - (7)(-4)}{42 + 4} = \frac{280}{46} = 6.09$

45. $v_1 = \frac{4\pi r^3}{3}$, $v_2 = \frac{4\pi(r+1)^3}{3}$

$v_2 - v_1 = \frac{4\pi(r+1)^3}{3} - \frac{4\pi r^3}{3} = 15$

$\frac{4\pi(r^3 + 3r^2 + 3r + 1)}{3} - \frac{4\pi r^3}{3} - 15 = 0$

$4\pi\left(r^2 + r + \frac{1}{3}\right) - 15 = 0$

$4\pi\left(r^2 + r + \frac{1}{3} - \frac{15}{4\pi}\right) = 0$

$r = \frac{-1 \pm \sqrt{1 - 4\left(\frac{1}{3} - \frac{15}{4\pi}\right)}}{2}$,   negative root cannot be a radius

$r = \frac{-1 + \sqrt{1 + 3.441}}{2} = .553$

Chapter 13 Sequences, Probability and Mathematical Induction

Section 13.1   Sequences in General

1. $a_n = n + 1$

   $a_1 = 1 + 1 = 2, \quad a_2 = 2 + 1 = 3, \quad a_3 = 3 + 1 = 4$

   $a_4 = 4 + 1 = 5, \quad a_5 = 5 + 1 = 6$

3. $a_n = 1 + (-1)^n$

   $a_1 = 1 - (-1)^1 = 2, \quad a_2 = 1 - (-1)^2 = 0, \quad a_3 = 1 - (-1)^3 = 2,$

   $a_4 = 1 - (-1)^4 = 0, \quad a_5 = 1 - (-1)^5 = 2$

5. $a_n = \cos n\pi$

   $a_1 = \cos \pi = -1, \quad a_2 = \cos 2\pi = 1, \quad a_3 = \cos 3\pi = -1,$

   $a_4 = \cos 4\pi = 1, \quad a_5 = \cos 5\pi = -1$

7. $a_n = \dfrac{n + 2}{n + 1}$

   $a_1 = \dfrac{1 + 2}{1 + 1} = \dfrac{3}{2}, \quad a_2 = \dfrac{2 + 2}{2 + 1} = \dfrac{4}{3}, \quad a_3 = \dfrac{3 + 2}{3 + 1} = \dfrac{5}{4},$

   $a_4 = \dfrac{4 + 2}{4 + 1} = \dfrac{6}{5}, \quad a_5 = \dfrac{5 + 2}{5 + 1} = \dfrac{7}{6}$

9. $a_n = n^2$

   $a_1 = 1^2 = 1, \quad a_2 = 2^2 = 4, \quad a_3 = 3^2 = 9, \quad a_4 = 4^2 = 16 \quad u_5 = 5^2 = 25$

11. $a_n = 2n + 1$

   $a_1 = 2(1) + 1 = 3, \quad a_2 = 2(2) + 1 = 5, \quad a_3 = 2(3) + 1 = 7,$

   $a_4 = 2(4) + 1 = 9, \quad a_5 = 2(5) + 1 = 11$

13. $a_n = \dfrac{1}{n^2}$

   $a_1 = \dfrac{1}{1} = 1, \quad a_2 = \dfrac{1}{2^2} = \dfrac{1}{4}, \quad a_3 = \dfrac{1}{3^2} = \dfrac{1}{9}, \quad a_4 = \dfrac{1}{4^2} = \dfrac{1}{16}, \quad a_5 = \dfrac{1}{5^2} = \dfrac{1}{25}$

15. $a_n = 5\left(\frac{1}{2}\right)^n$

$$a_1 = 5\left(\frac{1}{2}\right)^1 = \frac{5}{2}, \quad a_2 = 5\left(\frac{1}{2}\right)^2 = \frac{5}{4}, \quad a_3 = 5\left(\frac{1}{2}\right)^3 = \frac{5}{8},$$

$$a_4 = 5\left(\frac{1}{2}\right)^4 = \frac{5}{16}, \quad a_5 = 5\left(\frac{1}{2}\right)^5 = \frac{5}{32}$$

27. 1st year: $4000 - 4000(.25) = 4000(.75)$

2nd year: $4000(.75)^2$

3rd year: $4000(.75)^3$

4th year: $4000(.75)^4$

Section 13.2   Sequences of Partial Sums

1. $\displaystyle\sum_{m=1}^{5} \frac{1}{m+2} = \frac{1}{3} + \frac{1}{4} + \frac{1}{5} + \frac{1}{6} + \frac{1}{7} = 1.093$

3. $\displaystyle\sum_{n=0}^{5} n^2 = 0 + 1^2 + 2^2 + 3^2 + 4^2 + 5^2 = 55$

5. $\displaystyle\sum_{j=1}^{5} 3j = 3\sum_{j=1}^{5} = 3(1+2+3+4+5) = 45$

7. $\displaystyle\sum_{k=1}^{5}(1+k) = \sum_{k=1}^{5} 1 + \sum_{k=1}^{5} = 5 + (1+2+3+4+5) = 20$

9. $\displaystyle\sum_{n=-1}^{2} n^2 = (-2)^2 + (-1)^2 + 0 + 1^2 + 2^2 = 10$

19. $\displaystyle\sum_{i=1}^{n}(x_i + d) = (x_1 + d) + (x_2 + d) + \ldots + (x_n + d)$

$$= x_1 + x_2 + \ldots + x_n + nd$$

$$= \sum_{i=1}^{n} x_i + nd, \qquad \text{True}$$

21. $\displaystyle\sum_{k=2}^{5} a_{k-2} = a_0 + a_1 + a_2 + a_3 = \sum_{k=0}^{3} a_k, \qquad \text{True}$

23. $\displaystyle\sum_{n=1}^{3} x^n = x^1 + x^2 + x^3$

$\displaystyle\sum_{t=2}^{4} x^{t-1} = x^1 + x^2 + x^3,$ True

27. $S_n = (-1)^n$

$S_1 = a_1 = -1$

$S_2 = S_1 + a_2 = 1$

    $-1 + a_2 = 1, \quad a_2 = 2$

$S_3 = S_2 + a_3 = -1$

    $1 + a_3 = -1, \quad a_3 = -2$

$S_4 = S_3 + a_4 = 1$

    $-1 + a_4 = 1, \quad a_4 = 2$

So $a_1 = -1, \quad a_n = 2(-1)^n$

Section 13.3   Arithmetic Sequences

1. $2-1=1,\ 3-2=1,\ d=1$

3. $4-2=2,\ 6-4=2,\ d=2$

5. $6-3=3,\ 9-6=3,\ d=3$

7. $\frac{1}{2}-1=-\frac{1}{2},\ \frac{1}{4}-\frac{1}{2}=-\frac{1}{4},$ No

9. $\frac{1}{3}-\frac{1}{2}=-\frac{1}{6},\ \frac{1}{4}-\frac{1}{3}=-\frac{1}{12},$ No

11. $a_1 = 2,\ d = 3$

    $2,\ 5,\ 8,\ 11\ldots$

    $a_n = 2 + (n-1)3$

        $= 3n-1$

13. $S_n = a_1 n + \dfrac{n(n-1)d}{2}$

    $35 = 2(7) + \dfrac{7(6)d}{2},\ d = 1$

15. $a_1 = a_2 - d$

$a_1 = 5 - (-4) = 9$

$S_{10} = 9(10) + \dfrac{10(9)(-4)}{2}$

$= -90$

17. $\displaystyle\sum_{k=1}^{8}(2k-3)$

$a_1 = -1,\ a_2 = 1,\ d = 1 - (-1) = 2$

$S_8 = (-1)(8) + \dfrac{(8)(7)(2)}{2} = 48$

19. $\displaystyle\sum_{k=1}^{4}(7k-4)$

$a_0 = -4,\ a_1 = 3,\ d = 3 - (-4) = 7$

$S_5 = -4(5) + \dfrac{(5)(4)(7)}{2} = 50$

21. $\displaystyle\sum_{k=0}^{4}7(k-4)$

$a_0 = -28,\ a_1 = -21,$

$d = -21 - (-28) = 7$

$S_5 = -28(5) + \dfrac{(5)(4)(7)}{2} = -70$

23. $\displaystyle\sum_{i=1}^{n}(2n-1)$

$a_1 = 1,\ d = 2$

$S_n = 1(n) + \dfrac{(n)(n-1)2}{2}$

$= n + n(n-1) = n^2$

25. Let $a_1 = 1,\ a_2 = 4,\ a_3 = 9$

$d = a_2 - a_1 = 4 - 1 = 3$

$S_3 = 1(3) + \dfrac{3(2)3}{2} = 12$

or

$d = a_3 - a_2 = 9 - 4 = 5$

$S_3 = 1(3) = \dfrac{3(2)5}{2} = 18$

Actual $S_3 = 1 + 4 + 9 = 14$

27. $f(x) = x + 2$

$f(1) = 1+2 = 3,\ f(2) = 2+2 = 4,\ f(3) = 3+2 = 5$

$f(4) = 4+2 = 6,\ f(5) = 5+2 = 7$

29. $f(x) = -x+2$

$f(1) = -1+2 = 1,\ f(2) = -2+2 = 0,\ f(3) = -3+2 = -1$

$f(4) = -4+2 = -2,\ f(5) = -5+2 = -3$

31. $f(x) = 2x+5$

$f(1) = 2(1)+5 = 7,\ f(2) = 2(2)+5 = 9,\ f(3) = 2(3)+5 = 11$

$f(4) = 2(4)+5 = 13,\ f(5) = 2(5)+5 = 15$

33. $a_1 = 3$, $d = 6-3 = 3$

$f(x) = 3 + (x-1)3 = 3x$

35. $a_1 = 4$, $d = 3$

$f(x) = 4 + (x-1)3 = 3x+1$

37. $a_1 = 5$, $a_8 = 26$

$a_n = a_1 + (n-1)d$

$a_8 = 26 = 5+7d$, $d = 3$

$f(x) = 5 + (x-1)3 = 3x+2$

Section 13.4   Geometric Sequences

1. $r = \dfrac{a_n}{a_n-1} = \dfrac{12}{4} = 3$

$r = \dfrac{36}{12} = 3$,   yes, $a_n = 4(3)^{n-1}$

3. $r = \dfrac{6}{3} = 3$

$r = \dfrac{9}{6} = \dfrac{3}{2}$,   No

5. $r = \dfrac{-\frac{2}{3}}{\frac{1}{3}} = -2$

$r = \dfrac{\frac{4}{3}}{-\frac{2}{3}} = -2$,   yes, $a_n = \left(\dfrac{1}{3}\right)(-2)^{n-1}$

7. $r = \dfrac{\frac{1}{4}}{\frac{1}{2}} = \dfrac{1}{2}$

$r = \dfrac{\frac{1}{8}}{\frac{1}{4}} = \dfrac{1}{2}$,   yes, $a_n = \dfrac{1}{2}\left(\dfrac{1}{2}\right)^{n-1} = \left(\dfrac{1}{2}\right)^{n}$

9. $a_1 = 3$, $r = 5$

$a_4 = 3(5)^3 = 375$

$s_4 = \dfrac{1-5^4}{1-5}(3) = \dfrac{-624}{-4}(3) = 468$

11. $a_1 = 5$, $a_5 = 80$

$a_5 = 80 = 5r^4$

$r = \pm 2$

$r = 2$, $S_5 = \dfrac{(1-2^5)}{1-2}(5) = 155$

$r = -2$, $S_5 = \dfrac{\left(1-(-2)^5\right)(5)}{1-(-2)} = 55$

13. $a_1 = -2, \quad r = \frac{2}{3}$

$a_{10} = -2\left(\frac{2}{3}\right)^9 = \frac{-1024}{19{,}683}$

15. $\sum_{n=1}^{5} 2(3)^n, \quad a_1 = 6, \quad r = 3$

$S_5 = \frac{(1-3^5)6}{1-3} = 726$

17. $\sum_{k=1}^{10}\left[\left(\frac{1}{2}\right)^k + \left(\frac{3}{2}\right)^k\right]$

$S_{10} = \frac{\left(1-(\frac{1}{2})^{10}\right)\left(\frac{1}{2}\right)}{1-\frac{1}{2}} + \frac{\left(1-(\frac{3}{2})^{10}\right)\left(\frac{3}{2}\right)}{1-\frac{3}{2}}$

$= \left(1-\frac{1}{2^{10}}\right) + \left(1-\frac{3^{10}}{2^{10}}\right)(-3)$

$= \frac{3^{11}-1}{2^{10}} - 2$

19. $\sum_{k=1}^{6} 5\left(\frac{1}{2}\right)^k, \quad a_1 = \frac{5}{2}, \quad r = \frac{1}{2}$

$S_6 = \left(\frac{1-\left(\frac{1}{2}\right)^6}{1-\frac{1}{2}}\right)\left(\frac{5}{2}\right) = \left(\frac{1-\frac{1}{2^6}}{\frac{1}{2}}\right)\frac{5}{2}$

$= \left(1-\frac{1}{2^6}\right)5 = \frac{315}{64}$

21. $\sum_{n=1}^{3} \left(\frac{1}{4}\right)^n, \quad a_1 = \frac{1}{4}, \quad r = \frac{1}{4}$

$S_3 = \left(\frac{1-\left(\frac{1}{4}\right)^3}{1-\frac{1}{4}}\right)\left(\frac{1}{4}\right) = \left(\frac{1-\frac{1}{64}}{\frac{3}{4}}\right)\frac{1}{4} = \frac{21}{64}$

23. $r = \frac{4}{5}, \quad a_n = \left(\frac{4}{5}\right)^n a_1$

$\frac{1}{4} a_1 = \frac{4^n}{5^n}$

$5^n = 4^{n+1}$

$n \ln 5 = (n+1) \ln 4$

$n (\ln 5 - \ln 4) = \ln 4$

$n = \dfrac{\ln 4}{\ln 5 - \ln 4} = 6.212 \Rightarrow 7$ strokes

25. $a_3 = 4, \quad a_4 = 8$

$r = \frac{8}{4} = 2$

$a_n = a_1 2^n$

$a_3 = 4 = a_1 2^3$

$a_1 = \frac{1}{2}$

27. $f(x) = 3^x$

$f(1) = 3, \quad f(2) = 3^2 = 9, \quad f(3) = 3^3 = 27, \quad f(4) = 3^4 = 81, \quad f(5) = 3^5 = 243$

29. $f(z) = \left(\frac{1}{2}\right)^x$

$f(1) = \frac{1}{2}, \quad f(2) = \left(\frac{1}{2}\right)^2 = \frac{1}{4}, \quad f(3) = \left(\frac{1}{2}\right)^3 = \frac{1}{8}, \quad f(4) = \left(\frac{1}{2}\right)^4 = \frac{1}{16}, \quad f(5) = \left(\frac{1}{2}\right)^5 = \frac{1}{32}$

33. $p = 8000$

$t = 12{,}000$ after 6 years

$t = p(i+r)^6$

$12000 = 8000(i+r)^6$

$1.5 = (i+r)^6$

$i+r = (1.5)^{1/6}$

$r = (1.5)^{1/6} - 1 = .07 = 7\%$

35. Let w_0 be initial weight

$.10w = w_0 \left(\frac{1}{2}\right)^n$

$\frac{w_0}{10} = w_0 \left(\frac{1}{2}\right)^n$

$\frac{1}{10} = \left(\frac{1}{2}\right)^n$

$\ln .1 = n \ln .5$

$n = 3.32$

Section 13.5   The General Power of a Binomial

1. $(x - y)^6 = x^6 - 6x^5 y + 15x^4 y^2 - 20x^3 y^3 + 15x^2 y^4 - 6xy^5 + y^6$

3. $(3x - y)^5 = (3x)^5 - 5(3x)^4 y + 10(3x)^3 y^2 - 10(3x)^2 y^3 + 5(3x)y^4 - y^5$

$= 243x^5 - 405x^4 y + 270x^3 y^2 - 90x^2 y^3 + 15xy^4 - y^5$

5. $\left(x + \frac{1}{x}\right)^5 = x^5 + 5x^4\left(\frac{1}{x}\right) + 10x^3\left(\frac{1}{x}\right)^2 + 10x^2\left(\frac{1}{x}\right)^3 + 5x\left(\frac{1}{x}\right)^4 + \left(\frac{1}{x}\right)^5$

$$= x^5 + 5x^3 + 10x + \frac{10}{x} + \frac{5}{x^3} + \frac{1}{x^5}$$

7. $(y^2 + 3x)^3 = (y^2)^3 + 3(y^2)^2(3x) + 3y^2(3x)^2 + (3x)^3$

$$= y^6 + 9y^4x + 27y^2x^2 + 27x^3$$

9. $(\sqrt{x} + \sqrt{y})^6 = (\sqrt{x})^6 + 6(\sqrt{x})^5\sqrt{y} + 15(\sqrt{x})^4(\sqrt{y})^2 + 20(\sqrt{x})^3(\sqrt{y})^3$

$$+ 15(\sqrt{x})^2(\sqrt{y})^4 + 6(\sqrt{x})(\sqrt{y})^5 + (\sqrt{y})^6$$

$$= x^3 + 6x^2\sqrt{xy} + 15x^2y + 20xy\sqrt{xy} + 15xy^2 + 6y^2\sqrt{xy} + y^3$$

11. $\left(x^{1/3} + x^{-3/4}\right)^4 = \left(x^{1/3}\right)^4 + 4\left(x^{1/3}\right)^3 x^{-3/4} + 6\left(x^{1/3}\right)^2\left(x^{-3/4}\right)^2$

$$+ 4\left(x^{1/3}\right)\left(x^{-3/4}\right)^3 x^{-3/4} + \left(x^{-3/4}\right)^4$$

$$= x^{4/3} + 4x^{1/4} + 6x^{-5/6} + 4x^{-23/12} + x^{-3}$$

13. $(2+.1)^4 = 2^4 + 4(2)^3(.1) + 6(2)^2(.1)^2 + 4(2)(.1)^3 + (.1)^4$

$$= 16 + 3.2 + .24 + .008 + .0001 = 19.4418$$

15. $\left(x^{-3/4} - x^{3/4}\right)^3 = \left(x^{-3/4}\right)^3 - 3\left(x^{-3/4}\right)^2 x^{3/4} + 3\left(x^{-3/4}\right)\left(x^{3/4}\right)^2 - \left(x^{3/4}\right)^3$

$$= x^{-9/4} - 3x^{-3/4} + 3x^{3/4} - x^{9/4}$$

17. $(\sqrt{x} + \sqrt{y})^{10}$

4th term $= C_{10,\,3}(\sqrt{x})^7(\sqrt{y})^3$

$$= \frac{10!}{3!\,7!}x^3\sqrt{x}\,y\sqrt{y} = 120x^3y\sqrt{xy}$$

19. $\left(3x^3 - 7y^{1/3}\right)^{20}$

$\qquad$ 12th term $= C_{20,\,11}(3x^3)^9\left(-7y^{1/3}\right)^{11}$

$\qquad\qquad = \dfrac{20!}{9!\,11!}3^9 x^{27}(-1)7^{11}y^{11/3}$

$\qquad\qquad = \dfrac{-20!\ 3^9 7^{11}}{9!\,11!}x^{27}y^{11/3}$

21. $(x^3 - 2y^2)^5$

$\qquad Ax^{12}y^2 = C_{5,\,k}(x^3)^{5-k}(-2y^2)^k$

$\qquad 12 = 3(5-k), \quad k = 1$

$\qquad Ax^{12}y^2 = C_{5,\,1}(x^3)^4(-2y^2)^1$

$\qquad Ax^{12}y^2 = \dfrac{5!}{4!\,1!}x^{12}(-1)2y^2$

$\qquad\quad A = \dfrac{-2 \cdot 5!}{4!} = -10$

23. $\left(x + \dfrac{1}{3x}\right)^8$

$\qquad$ 5th term $= C_{8,\,4}x^4\left(\dfrac{1}{3x}\right)^4$

$\qquad\qquad = \dfrac{8!}{4!\,4!}x^4 \cdot \dfrac{1}{81x^4} = \dfrac{70}{81}$

25. $(2.3)^6 = (2+.3)^6 \simeq 2^6 + 6(2^5)(.3) = 121.66$

27. $(4 - .01)^3 \simeq 4^3 - 3(4)^2(.01) = 63.52$

29. $(1.01)^{500} = (1+.01)^{500} = 1^{500} + C_{500,\,1}1^{499}(.01) = 1 + 500(.01) = 6$

31. $\sqrt{24} = (25-1)^{1/2} = 25^{1/2}\left(1 - \dfrac{1}{25}\right)^{1/2}$

$\qquad = 5(1-.04)^{1/2} \simeq 5\left[1 + \dfrac{1}{2}(-.04) + \dfrac{\frac{1}{2}(-\frac{1}{2})}{1(2)}(-.04)^2\right] = 4.899$

33. $\sqrt[3]{7} = (8-1)^{1/3} = 8^{1/3}\left(1-\dfrac{1}{8}\right)^{1/3} = 2(1-.125)^{1/3}$

$\qquad\qquad \simeq 2\left[1 + \dfrac{1}{3}(-.125) + \dfrac{\left(\frac{1}{3}\right)\left(-\frac{2}{3}\right)(-.125)^2}{1(2)}\right] = 1.913194$

35. $\sqrt[3]{30} = (27+3)^{1/3} = 27^{1/3}\left(1+\frac{1}{9}\right)^{1/3} = 3(1+.1111)^{1/3}$

$$\simeq 3\left[1 + \frac{1}{3}(.1111) + \frac{\left(\frac{1}{3}\right)\left(-\frac{2}{3}\right)(.1111)^2}{(1)(2)}\right] = 3.107$$

37. $\sqrt[4]{80} = (81-1)^{1/4} = 81^{1/4}\left(1-\frac{1}{81}\right)^{1/4} = 3(1-.01235)^{1/4}$

$$\simeq 3\left[1 + \frac{1}{4}(-.01235) + \frac{\left(\frac{1}{4}\right)\left(-\frac{3}{4}\right)(-.01235)^2}{(2)(1)}\right] = 2.9907$$

39. $(98)^{1/2} = (100-2)^{1/2} = 100^{1/2}\left(1-\frac{2}{100}\right)^{1/2} = 10(1-.02)^{1/2}$

$$\simeq 10\left[1 + \frac{1}{2}(-.02) + \frac{\frac{1}{2}\left(-\frac{1}{2}\right)(-.02)^2}{(1)(2)}\right] = 9.8995$$

Section 13.6   Permutations and Combinations

1. $P_{7,\,4} = \frac{7!}{3!} = 7 \cdot 6 \cdot 5 \cdot 4 = 840$

$C_{7,\,4} = \frac{7!}{4!\,3!} = 35$

3. $P_{n,\,3} = 9P_{n,\,2}$

$\frac{n!}{(n-3)!} = \frac{9n!}{(n-2)!}$

$\frac{1}{(n-3)!} = \frac{9}{(n-2)!}$

$(n-2)! = 9(n-3)!$

$n-2 = 9$

$n = 11$

5. $\displaystyle\sum_{k=1}^{4} P_{4,\,k}$

$= \frac{4!}{3!} + \frac{4!}{2!} + \frac{4!}{1!} + \frac{4!}{0!}$

$= 4 + 4(3) + 4(3)(2) + 4(3)(2)$

$= 64 = 4^3$

7. $C_{8,\,3} = \frac{8!}{3!\,5!} = 56$

9. $P_{4,\,4} = \frac{4!}{0!} = 24$

11. $9 \cdot 8 \cdot 7 \cdot 6 \cdot 5 \cdot 4 \cdot 3 = 181,440$

13. $\dfrac{13 \cdot 4!}{3! \ 1!} \cdot \dfrac{12 \cdot 4!}{2! \ 2!} = 13 \cdot 4 \cdot 12 \cdot 3 \cdot 2 = 3744$

15. $5 \cdot 5 \cdot 5 \cdot 2 = 250$

17. $C_{10,\ 4} + C_{8,\ 2} = \dfrac{10!}{6! \ 4!} - \dfrac{8!}{6! \ 2!} = 210 - 28 = 182$

Section 13.7   Probability

1. $P\,(\text{a heart}) = \dfrac{13}{52} = \dfrac{1}{4}$

3. $P\,(\text{bolt}) = \dfrac{3}{8}$

   $P\,(\text{nut}) = \dfrac{5}{8}$

5. $P\,(\text{odd}) = \dfrac{3}{6} = \dfrac{1}{2}$

7. $P\,(\text{2 heads}) = \dfrac{C_{3,\ 2}}{8} = \dfrac{3!}{2! \ 1!} \cdot \dfrac{1}{8} = \dfrac{3}{8}$

9. $P\,(\text{not red}) = \dfrac{C_{10,\ 1}}{C_{13,\ 1}} = \dfrac{10!}{1! \ 9!} \cdot \dfrac{12! \ 1!}{13!} = \dfrac{10}{13}$

11. $P\,(\text{3 tails}) = \dfrac{C_{6,\ 3}}{2^6} = \dfrac{6!}{3! \ 3!} \cdot \dfrac{1}{64} = \dfrac{5}{16}$

13. $P\,(\text{2 nickels}) = \dfrac{C_{15,\ 2}}{C_{26,\ 2}} = \dfrac{15!}{13! \ 2!} \cdot \dfrac{24! \ 2!}{26!} = \dfrac{21}{65}$

15. $P\,(\text{vowels}) = \dfrac{C_{14,\ 2}}{C_{11,\ 2}} = \dfrac{4!}{2! \ 2!} \cdot \dfrac{9! \ 2!}{11!} = \dfrac{6}{55}$

17. $P(3) = \dfrac{2}{36} = \dfrac{1}{18}$

19. $P\,(\text{exactly two acres}) = \dfrac{C_{4,\ 2}\ C_{48,\ 3}}{C_{52,\ 5}} = \dfrac{4!}{2! \ 2!} \cdot \dfrac{48!}{45! \ 3!} \cdot \dfrac{47! \ 5!}{52!} = \dfrac{2162}{54145}$

1. $\displaystyle\sum_{k=1}^{n} 2k = n(n+1)$

     Let $k = 1$, $2(1) = 1(1+1)$

                     $2 = 2 \ \checkmark$

     $S_k = k(k+1)$

     $S_{k+1} = S_k + 2(k+1)$

           $= k(k+1) + 2(k+1)$

           $= (k+1)(k+2)$

3. $\displaystyle\sum_{k=1}^{n} (3k-2) = \dfrac{n(3-1)}{2}$

     Let $k = 1$, $3-2 = \dfrac{1(3-1)}{2}$

                      $1 = 1 \ \checkmark$

     $S_k = \dfrac{k(3k-1)}{2}$

     $S_{k+1} = S_k + \left[3(k+1) - 2 \right]$

           $= \dfrac{k(3k-1)}{2} + 3(k+1) - 2$

           $= \dfrac{3k^2 + 5k + 2}{2}$

           $= \dfrac{(3k+2)(k+1)}{2}$

           $= \dfrac{\left[3(k+1) - 1 \right](k+1)}{2}$

5. $\displaystyle\sum_{k=1}^{n} k(k+1) = \dfrac{n(n+1)(n+1)}{3}$

     Let $k = 1$, $1(2) = \dfrac{1(2)(3)}{3}$

                      $2 = 2 \ \checkmark$

     $S_k = \dfrac{k(k+1)(k+2)}{3}$

     $S_{k+1} = S_k + (k+1)(k+2)$

           $= \dfrac{k(k+1)(k+2)}{3} + (k+1)(k+2)$

           $= (k+1)(k+2)(\tfrac{k}{3}+1)$

           $= \dfrac{(k+1)(k+2)(k+3)}{3}$

7. $\sum_{k=1}^{n} k^2 = \dfrac{n(n+1)(2n+1)}{6}$

Let $k = 1$, $\quad 1 = \dfrac{1(2)(3)}{6}$

$$1 = 1 \checkmark$$

$S_k = \dfrac{k(k+1)(2k+1)}{6}$

$S_{k+1} = S_k + (k+1)^2$

$\quad = \dfrac{k(k+1)(2k+1)}{6} + (k+1)^2$

$\quad = \dfrac{(k+1)\big(k(2k+1) + 6(k+1)\big)}{6}$

$\quad = \dfrac{(k+1)(2k^2 - 7k + 6)}{6}$

$\quad = \dfrac{(k+1)(2k+3)(k+2)}{6}$

$\quad = \dfrac{(k+1)(k+2)\big(2(k+1) + 1\big)}{6}$

9. $\sum_{k=1}^{n} r^{k-1} = \dfrac{1-r^n}{1-r}, \qquad r \neq 1$

Let $k = 1$, $\quad r^\circ = \dfrac{1-r}{1-r}$

$$1 = 1 \checkmark$$

$S_k = \dfrac{1-r^k}{1-r}$

$S_{k+1} = S_k + r^{(k+1)-1}$

$\quad = \dfrac{1-r^k}{1-r} + r^k$

$\quad = \dfrac{1-r^k + r^k(1-r)}{1-r}$

$\quad = \dfrac{1-r^k + r^k - r^{k+1}}{1-r}$

$\quad = \dfrac{1-r^{k+1}}{1-r}$

11. $n(n+1)(n+2) = 6c,\quad c$ is an integer

Let $n = 1$, $1(2)(3) = 6(1)$

$$6 = 6 \checkmark$$

Assume true for k.

$\quad k(k+1)(k+2) = 6c$, for some integer c

Consider $(k+1)(k+2)(k+3)$

$\quad = k^3 + 6k^2 + 11k + 6$

$\quad = k^3 + 3k^2 + 3k^2 + 9k + 2k + 6$

$\quad = (k^3 + 3k^2 + 2k) + (3k^2 + 9k + 6)$

$\quad = k(k^2 + 3k + 2) + 3(k^2 + 3k + 2)$

$\quad = k(k+1)(k+2) + 3(k+1)(k+2)$

Let $k = $ odd number

Then $k+1 = $ even

$\quad k+1 = 2d$ for some integer d

So $(k+1)(k+2)(k+3) = 6c + 3(2d)(k+2)$

$\quad\quad\quad = 6(c+d)(k+2),$

c, d, $k+1$ are integers

so $(c+d)(k+1)$ is an integer

Let $k =$ even number

Then $k+2 =$ even

$k+2 = 2d$ for some integer d

So $(k+1)(k+2)(k+3) = 6c + 3(2d)(k+1)$

$= 6(c+d)(k+1)$

c, d, $k+1$ are integers

so $(c+d)(k+1)$ is an integer

13. $\displaystyle\sum_{k=1}^{n+1} \frac{1}{n+k} \leq \frac{5}{6}$

Let $n = 1$, $\displaystyle\sum_{k=1}^{2} \frac{1}{1+k} = \frac{1}{1+1} + \frac{1}{1+2} \leq \frac{5}{6}$

Assume true for $n = r$

$n = r = \displaystyle\sum_{k=1}^{r+1} \frac{1}{r+k} = \frac{1}{r+1} + \frac{1}{r+2} + \cdots + \frac{1}{r+r+1} \leq \frac{5}{6}$

Consider $\displaystyle\sum_{k=1}^{(r+1)+1} \frac{1}{(r+1)+k} = \sum_{k=1}^{r+2} \frac{1}{r+1+k}$

$= \dfrac{1}{r+2} + \dfrac{1}{r+3} + \cdots + \dfrac{1}{(r+1)+(r+2)}$

$= \dfrac{1}{r+2} + \dfrac{1}{r+3} + \cdots + \dfrac{1}{2r+3}$

$= \dfrac{1}{r+2} + \dfrac{1}{r+3} + \cdots + \dfrac{1}{2r+1} + \left(\dfrac{1}{2r+2} + \dfrac{1}{2r+3}\right)$

$\leq \dfrac{1}{r+2} + \dfrac{1}{r+3} + \cdots + \dfrac{1}{2r+1} + \dfrac{1}{r+1}$

$= \displaystyle\sum_{k=1}^{r+1} \frac{1}{r+k} \leq \frac{5}{6}$

11. $a_1 = 1, \quad d = 2$

   $S_{20} = 1(20) + \dfrac{20(19)(2)}{2} = 400$

13. $a_1 = 3, \quad d = 6$

   $S_{20} = 3(20) + \dfrac{20(19)(6)}{2} = 1200$

15. $a_1 = 1, \quad d = 0$

   $S_{20} = 1(20) + 0 = 20$

17. $a_1 = 2, \quad r = 2$

   $S_{20} = \dfrac{(1-2^{20})2}{1-2} = (1-2^{20})(-2) = -2 + 2^{21} = 2{,}097{,}150$

19. $a_1 = -5, \quad r = -5$

   $S_{20} = \dfrac{\left(1 - (-5)^{20}\right)(-5)}{1 - (-5)} = \dfrac{(1-5^{20})(-5)}{6} = \dfrac{-5 + 5^{21}}{6} = 7.947 \times 10^{13}$

21. $\displaystyle\sum_{k=1}^{2} 2^k = 2 + 2^2 = 6$

23. $\displaystyle\sum_{i=1}^{4} i(i+2) = 1(3) + 2(4) + 3(5) + 4(6) = 50$

25. $\displaystyle\sum_{k=2}^{10}\left[\dfrac{1}{k+3} - \dfrac{1}{k+2}\right] = \left(\dfrac{1}{5} - \dfrac{1}{4}\right) + \left(\dfrac{1}{6} - \dfrac{1}{5}\right) + \left(\dfrac{1}{7} - \dfrac{1}{6}\right) + \left(\dfrac{1}{8} - \dfrac{1}{7}\right) + \left(\dfrac{1}{9} - \dfrac{1}{8}\right)$

   $+ \left(\dfrac{1}{10} - \dfrac{1}{9}\right) + \left(\dfrac{1}{11} - \dfrac{1}{10}\right) + \left(\dfrac{1}{12} - \dfrac{1}{11}\right) + \left(\dfrac{1}{13} - \dfrac{1}{12}\right)$

   $= -\dfrac{1}{4} + \dfrac{1}{13} = -\dfrac{9}{52}$

27. $\displaystyle\sum_{k=2}^{10} a_k = a_2 + a_3 + a_4 + a_5 + a_6 + a_7 + a_8 + a_9 + a_{10}$

   $\displaystyle\sum_{k=4}^{12} a_{k-2} = a_2 + a_3 + a_4 + a_5 + a_6 + a_7 + a_8 + a_9 + a_{10}, \quad$ True

29. $\displaystyle\sum_{t=1}^{5} (t^2-1)(t-2) = 0(-1) + (3)(0) + (8)(1) + (15)(2) + (24)(3) = 110$

   $\displaystyle\sum_{s=-1}^{3} (s^2-1)(s) = (2)(-1) + (1)(0) + (2)(1) + (5)(2) + (10)(3) = 40, \quad$ False

31. $\left(x - \frac{1}{x}\right)^{10}$

$$C_{10,\,k}x^k\left(-\frac{1}{x}\right)^{10-k} = C_{10,\,k}\frac{x^k}{x^{10-k}}(-1)^{10-k}$$

$$\frac{x^k}{x^{10-k}} = 1$$

$$x^{k-(10-k)} = x^{\circ}$$

$$k - (10-k) = 0$$

$$k = 5$$

$$C_{10,\,5}x^5\left(-\frac{1}{x}\right)^5 = \frac{10!}{5!\,5!}(-1)^5 = -252$$

33. $(x-5y)^4 = x^4 - 4x^3(5y) + 6x^2(5y)^2 - 4x(5y)^3 + (5y)^4$

$$= x^4 - 20x^3y + 150x^2y^2 - 500xy^3 + 625y^4$$

35. $(1.01)^{100} = (1+.01)^{100} \simeq 1^{100} + C_{100,\,1}1^{99}(.01) + C_{100,\,2}1^{98}(.01)^2$

$$= 1 + 100(.01) + \frac{100!}{98!\,2!}1(.01)^2 = 2.495$$

37. $2^{20} = 1,048,576$

39. $P_{5,\,2} = \frac{5!}{3!} = 20$

$$P_{100,\,2} = \frac{100!}{98!} = 9900$$

41. $\dfrac{(2n)!}{(2n-3)!} = \dfrac{(2n)(2n-1)(2n-2)\cancel{(2n-3)}\cancel{(2n-4)}\cdots\cancel{1}}{\cancel{(2n-3)}\cancel{(2n-4)}\cdots\cancel{1}}$

$$= 2n(2n-1)(2n-2)$$

47. $P\,(4 \text{ tails}) = \dfrac{C_{7,\,4}}{2^7} = \dfrac{7!}{4!\,3!} \cdot \dfrac{1}{2^7} = \dfrac{35}{128}$

Chapter 14 The Conic Sections

Section 14.1   Conic Sections: The Parabola

1. $y^2 = -8x$

   $4a = -8,\quad a = -2$

   Focus: $(-2, 0)$

   Directrix: $x = 2$

   Right Chord: $(-2, -4), (-2, 4)$

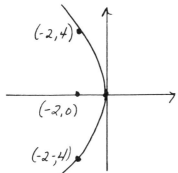

3. $2x^2 = 12y$

   $x^2 = 6y$

   $4a = 6,\quad a = \frac{3}{2}$

   Focus: $(0, \frac{3}{2})$

   Directrix: $y = -\frac{3}{2}$

   Right Chord: $(-3, \frac{3}{2}), (3, \frac{3}{2})$

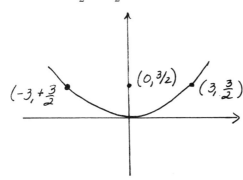

5. $y^2 + 16x = 0$

$y^2 = -16x$

$4a = -16, \quad a = -4$

Focus: $(-4, 0)$

Directrix: $x = 4$

Right Chord: $(-4, -8), (-4, 8)$

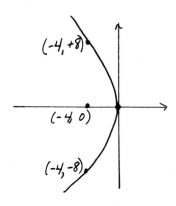

7. $y^2 = 3x$

$4a = 3, \quad a = \frac{3}{4}$

Focus: $\left(\frac{3}{4}, 0\right)$

Directrix: $x = -\frac{3}{4}$

Right Chord: $\left(\frac{3}{4}, \frac{3}{2}\right), \left(\frac{3}{4}, -\frac{3}{2}\right)$

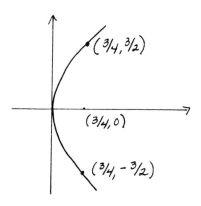

9. $y^2 = -2x$

$4a = -2, \quad a = -\frac{1}{2}$

Focus: $\left(-\frac{1}{2}, 0\right)$

Directrix: $x = \frac{1}{2}$

Right Chord: $\left(-\frac{1}{2}, -1\right), \left(-\frac{1}{2}, 1\right)$

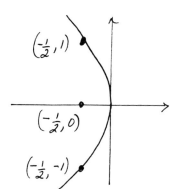

11. Focus: $(0, 2)$

   Directrix: $y = -2$

   $a = 2$

   $x^2 = 4ay, \quad x^2 = 8y$

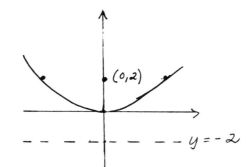

13. Focus: $\left(\frac{3}{2}, 0\right)$

   Directrix: $x = -\frac{3}{2}$

   $a = \frac{3}{2}$

   $y^2 = 4ax, \quad y^2 = 6x$

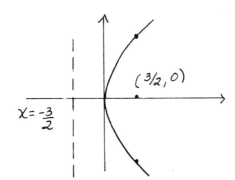

15. Right Chord: $(2, -1), (-2, -1)$

   Vertex: $(0, 0)$

   Focus: $(0, -1)$

   $a = -1$

   $x^2 = 4ay, \quad x^2 = -4y$

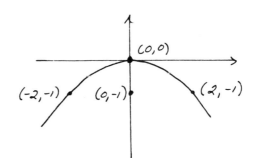

17. Vertex: (0, 0)

Vertical Axis $\rightarrow$ $x^2 = 4ay$

(2, 4)

$4 = 4a(4)$

$a = \frac{1}{4}$

$x^2 = 4\left(\frac{1}{4}\right)y = y$

19.

21.

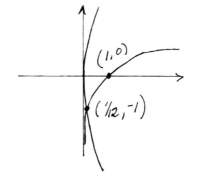

23. $x^2 = 4ay$

$250{,}000 = 4a(250)$

$4a = 1000$

$x^2 = 1000y$

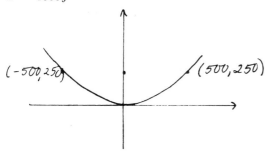

Section 14.2   The Ellipse and the Circle

1. $5x^2 + y^2 = 25$

$\dfrac{x^2}{5} + \dfrac{y^2}{25} = 1$

$a = 5, \quad b = \sqrt{5}$

$c = \sqrt{25-5} = \pm 2\sqrt{5}$

Focus: $(0, \pm 2\sqrt{5})$

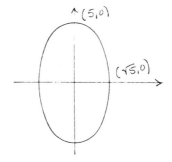

3. $16x^2 + 4y^2 = 16$

$x^2 + \dfrac{y^2}{4} = 1$

$a = 2, \quad b = 1$

$c = \sqrt{4-1} = \pm\sqrt{3}$

Focus: $(0, \pm\sqrt{3})$

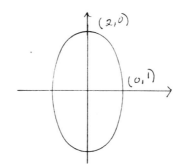

5. $x^2 + y^2 = 9$

   $r = 3$

7. $2x^2 + 3y^2 = 24$

   $\dfrac{x^2}{12} + \dfrac{y^2}{8} = 1$

   $a = 2\sqrt{3}, \quad b = 2\sqrt{2}$

   $c = \sqrt{12-8} = 2$

   Focus: $(\pm 2, 0)$

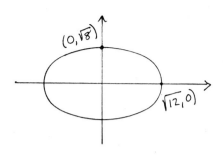

9. $9x^2 + 4y^2 = 4$

   $\dfrac{x^2}{\frac{4}{9}} + y^2 = 1$

   $a = 1, \quad b = \dfrac{2}{3}$

   $c = \sqrt{1-\dfrac{4}{9}} = \pm\dfrac{\sqrt{5}}{3}$

   Focus: $\left(0, \pm\dfrac{\sqrt{5}}{3}\right)$

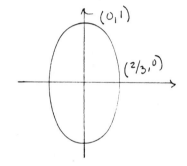

11. Vertices: $(\pm 4, 0)$

    Minor Axis: 6

    $\dfrac{x^2}{16} + \dfrac{y^2}{9} = 1$

    $9x^2 + 16y^2 = 144$

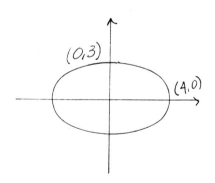

13. Vertices: $(0, \pm 5)$

Semi-minor Axis: $\frac{3}{2}$

$$\frac{x^2}{\frac{9}{4}} + \frac{y^2}{25} = 1$$

$$9y^2 + 100x^2 = 225$$

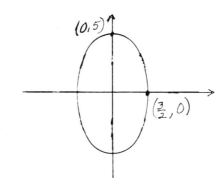

15. Major Axis: 10

Foci: $(\pm 4, 0)$

$a = 5, \quad c = 4$

$b^2 = 25 - 16 = 9, \quad b = 3$

$$\frac{x^2}{25} + \frac{y^2}{9} = 1$$

$$9x^2 + 25y^2 = 225$$

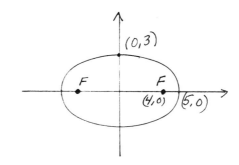

17. Foci: $(\pm 1, 0)$

Major Axis: 8

$c = 1, \quad a = 4$

$b^2 = 4^2 - 1 = 15, \quad b = \sqrt{15}$

$$\frac{x^2}{16} + \frac{y^2}{15} = 1$$

$$15x^2 + 16y^2 = 240$$

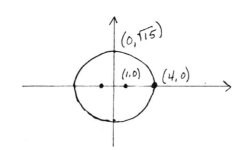

19. Vertices: $\left(\pm\frac{5}{2}, 0\right)$

$$\frac{x^2}{\frac{25}{4}} + \frac{y^2}{b^2} = 1$$

$$\frac{4x^2}{25} + \frac{y^2}{b^2} = 1$$

If $(1, 1)$, then

$$\frac{4}{25} + \frac{1}{b^2} = 1$$

$$b^2 = \frac{25}{21}, \quad b = \frac{5}{\sqrt{21}}$$

$$\frac{4x^2}{25} + \frac{21y^2}{25} = 1$$

$$4x^2 + 21y^2 = 25$$

21.

23.

25. Major Axis: 10

Minor Axis: 3

$$\frac{x^2}{25} + \frac{y^2}{\left(\frac{3}{2}\right)^2} = 1$$

$$9x^2 + 100y^2 = 225$$

Section 14.3   The Hyperbola

1. $x^2 - y^2 = 16$

     $\frac{x^2}{16} - \frac{y^2}{16} = 1$

     $a = 4, \quad b = 4$

     $c = \sqrt{16+16} = \sqrt{32}$

     Foci: $\left(+\sqrt{32}, 0\right)$

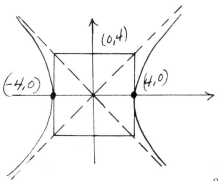

3. $4x^2 - 9y^2 = 36$

$$\frac{x^2}{9} - \frac{y^2}{4} = 1$$

$a = 3, \quad b = 2$

$c = \sqrt{9+4} = \sqrt{13}$

Foci: $\left(\pm\sqrt{13}, 0\right)$

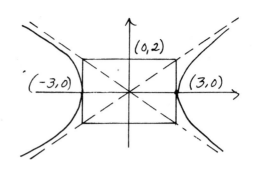

5. $4y^2 - 25x^2 = 100$

$$\frac{y^2}{25} - \frac{x^2}{4} = 1$$

$a = 5, \quad b = 2$

$c = \sqrt{25+4} = \sqrt{29}$

Foci: $\left(0, \pm\sqrt{29}\right)$

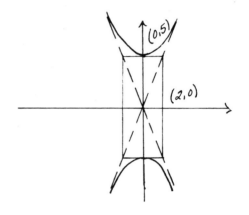

7. $4x^2 - 16y^2 = 25$

$$\frac{x^2}{\frac{25}{4}} - \frac{y^2}{\frac{25}{16}} = 1$$

$a = \frac{5}{2}, \quad b = \frac{5}{4}$

$c = \sqrt{\frac{25}{4} + \frac{25}{16}} = \frac{5}{4}\sqrt{5}$

Foci: $\left(\pm\frac{5}{4}\sqrt{5}, 0\right)$

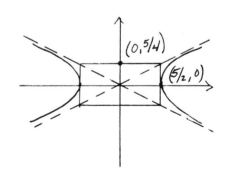

9. $y^2 + 1 = x^2$

$x^2 - y^2 = 1$

$a = 1, \quad b = 1$

$c = \sqrt{1+1} = \sqrt{2}$

Foci: $\left(\pm\sqrt{2}, 0\right)$

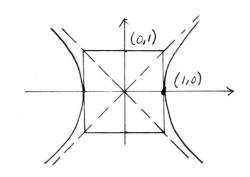

11. Vertices: $(\pm 4, 0)$

Foci: $(\pm 5, 0)$

$\dfrac{x^2}{a^2} - \dfrac{y^2}{b^2} = 1$

$a = 4, \quad c = 5$

$b = \sqrt{25-16} = 3$

$\dfrac{x^2}{16} - \dfrac{y^2}{9} = 1$

$9x^2 - 16y^2 = 144$

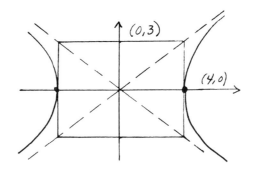

13. Conjugate Axis: 4

Vertices: $(0, \pm 1)$

$\dfrac{y^2}{a^2} - \dfrac{x^2}{b^2} = 1$

$a = 1, \quad b = 2$

$y^2 - \dfrac{x^2}{4} = 1$

$4y^2 - x^2 = 4$

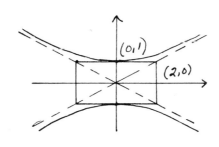

15. Transverse Axis: 6

Foci: $\left(\pm\frac{7}{2}, 0\right)$

$\frac{x^2}{a^2} - \frac{y^2}{b^2} = 1$

$a = 3, \quad c = \frac{7}{2}$

$b = \sqrt{\frac{49}{4} - 9} = \frac{\sqrt{13}}{2}$

$\frac{x^2}{9} - \frac{4y^2}{13} = 1$

$13x^2 - 36y^2 = 117$

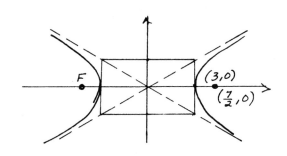

17. Vertices: $(0, \pm 4)$

Asymptotes: $y = \pm\frac{x}{2}$

$\frac{y^2}{a^2} - \frac{x^2}{b^2} = 1$

$a = 4$

$\frac{1}{2} = \frac{a}{b}, \quad b = 8$

$\frac{y^2}{16} - \frac{x^2}{64} = 1$

$4y^2 - 4x^2 = 64$

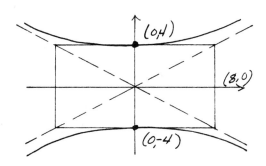

19. Vertices: $(0, \pm 3)$

$\frac{y^2}{a^2} - \frac{x^2}{b^2} = 1$

$a = 3$

$\frac{y^2}{9} - \frac{x^2}{b^2} = 1$

If $(2, 7)$, then

$\frac{49}{9} - \frac{4}{b^2} = 1, \quad b^2 = \frac{9}{10}$

$\frac{y^2}{9} - \frac{10x^2}{9} = 1$

$y^2 - 10x^2 = 9$

21. $\dfrac{x^2}{\sin^2 C} - \dfrac{y^2}{\cos^2 C} = 1$

    $a^2 = \sin^2 C$

    $b^2 = \cos^2 C$

    $c^2 = \sin^2 C + \cos^2 C$

    $c^2 = 1$

    $c = \pm 1$

    Foci: $(\pm 1, 0)$

Section 14.4   Translation of Axes

1. Vertex: $(3, 1)$
    Focus: $(5, 1)$

    $a = 5 - 3 = 2$

    $(y - k)^2 = 4a(x - h)$

    $(y - 1)^2 = 8(x - 3)$

    $8x = y^2 - 2y + 25$

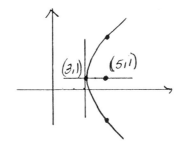

3. Directrix: $y = 2$
    Vertex: $(1, -1)$

    $a = -1 - 2 = -3$

    $(x - h)^2 = 4a(y - k)$

    $(x - 1)^2 = -12(y + 1)$

    $x^2 - 2x + 13 = -12y$

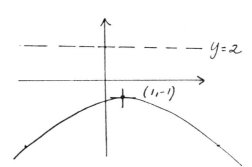

5. Right Chord: $(2, 4)$, $(2, 0)$

$4a = 4$, $a = 1$

Axis: $y = 2$

Vertix: $(1, 2)$

$(y-k)^2 = 4a(x-h)$

$(y-2)^2 = 4(x-1)$

$y^2 - 4y + 8 = 4x$

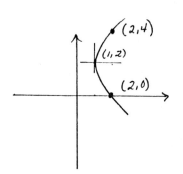

7. Major Axis: 8

Foci: $(5, 1)$, $(-1, 1)$

Center: $(2, 1)$

$a = 4$, $c = 3$,

$b = \sqrt{16-9} = \sqrt{17}$

$\dfrac{(x-h)^2}{a^2} + \dfrac{(y-k)^2}{b^2} = 1$

$\dfrac{(x-2)^2}{16} + \dfrac{(y-1)^2}{9} = 1$

$7x^2 - 28x + 16y^2 - 32y - 68 = 0$

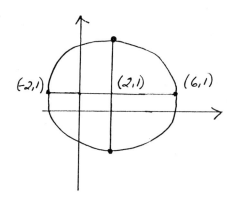

9. Minor Axis: 2

Vertices: $\left(\frac{1}{2}, 0\right)$, $\left(\frac{1}{2}, -8\right)$

Center: $\left(\frac{1}{2}, -4\right)$

$a = 4$, $b = 1$, $c = \sqrt{16+1} = \sqrt{17}$

$\dfrac{(x-h)^2}{b^2} + \dfrac{(y-k)^2}{a^2} = = 1$

$\left(x - \frac{1}{2}\right)^2 + \dfrac{(y+4)^2}{16} = 1$

$16x^2 - 16x + y^2 + 8y + 4 = 0$

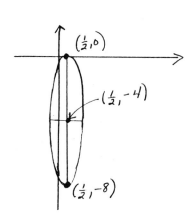

11. Vertices: $(-6, 3), (-2, 3)$

Foci: $(-5, 3), (-3, 3)$

Center: $(-4, 3)$

$a = 2, \quad c = 1, \quad b^2 = 4-1 = 3$

$\dfrac{(x-h)^2}{a^2} + \dfrac{(y-k)^2}{b^2} = 1$

$\dfrac{(x+4)^2}{4} + \dfrac{(y-3)^2}{3} = 1$

$3x^2 + 24x + 4y^2 - 24y + 72 = 0$

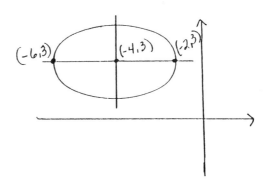

13. Center: $(-1, 2)$

Transverse Axis: 7

Conjugate Axis: 8

Vertical Axis

$a = \dfrac{7}{2} \qquad b = 4$

$\dfrac{(y-k)^2}{a^2} - \dfrac{(x-h)^2}{b^2} = 1$

$\dfrac{(y-2)^2}{\left(\frac{7}{2}\right)^2} - \dfrac{(x+1)^2}{4^2} = 1$

$64y^2 - 49x^2 - 98x - 256y - 577 = 0$

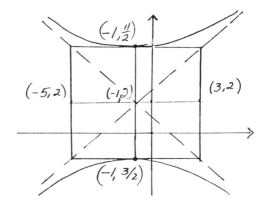

15. Vertices: $(5, 1), (-1, 1)$

Foci: $(6, 1), (-2, 1)$

Center: $(2, 1)$

$a = 3, \quad c = 4$

$b^2 = 16-9 = 7$

$\dfrac{(x-h)^2}{a^2} - \dfrac{(y-k)^2}{b^2} = 1$

$\dfrac{(x-2)^2}{9} - \dfrac{(y-1)^2}{7} = 1$

$7x^2 - 28x - 9y^2 + 18y - 44 = 0$

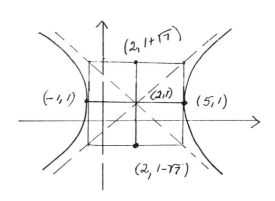

17. Vertices: $(-4, -2), (0, -2)$

Slope of Asymptotes: $\pm\frac{1}{2}$

Center: $(-2, -2)$

$a = 2, \qquad m = \pm\frac{1}{2}$

$\qquad\qquad m = \pm\frac{b}{a}, \quad b = 1$

$\dfrac{(x-h)^2}{a^2} - \dfrac{(y-k)^2}{b^2} = 1$

$\dfrac{(x+2)^2}{4} - \dfrac{(y+2)^2}{1} = 1$

$x^2 + 4x - 4y^2 - 16y - 16 = 0$

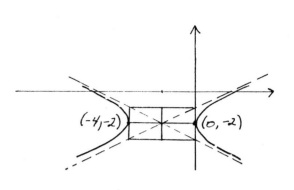

19. Center: $(h, 0)$

$(x-h)^2 + (y-0)^2 = r^2$

$(x-h)^2 + y^2 = r^2$

21. Vertex: $(h, 0)$

$(y-k)^2 = 4a(x-h)$

$y^2 = 4a(x-h)$

23. Center: $(h, 0)$

$(x-h)^2 + y^2 = r^2$

If $(0, 0)$, then

$(-h)^2 + 0 = r^2$

$h = r$

$(x-r)^2 + y^2 = r^2$

25.

$y = 20x - \dfrac{1}{10}x^2$

$10y = 200x - x^2$

$x^2 - 200x = -10y$

$x^2 - 200x + (100)^2 = -10y + (100)^2$

$(x-100)^2 = -10(y-1000)$

Section 14.5   Rotation of Axes

1. $x' = 3\cos 30° + 30° = \dfrac{3\sqrt{3} + 1}{2}$

$y' = -3\sin 30° + \cos 30° = \dfrac{-3 + \sqrt{3}}{2}$

3. $x' = -\cos 30° - \sin 30° = \dfrac{-\sqrt{3} - 1}{2}$

$y' = \sin 30° - \cos 30° = \dfrac{1 - \sqrt{3}}{2}$

5. $x' = 2 \cos 30° - \sin 30° = \dfrac{2\sqrt{3} - 1}{2}$

$y' = -2 \sin 30° - \cos 30° = \dfrac{-2 - \sqrt{3}}{2}$

7. $x = \dfrac{x'}{\sqrt{2}} - \dfrac{y'}{\sqrt{2}}$

$y = \dfrac{x'}{\sqrt{2}} + \dfrac{y'}{\sqrt{2}}$

$y = 2x$

$\dfrac{x' + y'}{\sqrt{2}} = \dfrac{2(x' - y')}{\sqrt{2}}$

$3y' = x'$

$y' = \frac{1}{3}x'$

9. $x = x' \cos 60° - y' \sin 60° = \dfrac{x'}{2} - \dfrac{\sqrt{3}\,y'}{2}$

$y = x' \sin 60° + y' \cos 60° = \dfrac{\sqrt{3}\,x'}{2} + \dfrac{y'}{2}$

$2x + 7y = 3$

$\dfrac{2\left(x' - \sqrt{3}\,y'\right)}{2} + \dfrac{7\left(\sqrt{3}\,x' + y'\right)}{2} = 3$

$\left(2 + 7\sqrt{3}\right)x' + \left(7 - 2\sqrt{3}\right)y' = 6$

11. $x = x' \cos 90° - y' \sin 90° = -y'$

$y = x' \sin 90° + y' \cos 90° = x'$

$x^2 + 4y^2 = 4$

$(-y')^2 + 4(x')^2 = 4$

$y'^2 + 4x'^2 = 4$

13. $x = x' \cos 30° - y' \sin 30° = \dfrac{\sqrt{3}x'}{2} - \dfrac{y'}{2}$

$y = x' \sin 30° + y' \cos 30° = \dfrac{x'}{2} + \dfrac{\sqrt{3}y'}{2}$

$$x^2 + 2x + y^2 = 0$$

$$\dfrac{\left(\sqrt{3}x' - y'\right)^2}{4} + \dfrac{2\left(\sqrt{3}x' - y'\right)}{2} + \dfrac{\left(x' + \sqrt{3}y'\right)^2}{4} = 0$$

$$\dfrac{3x'^2}{4} - \dfrac{\sqrt{3}x'y'}{2} + \dfrac{y'^2}{4} + \sqrt{3}x' - y' + \dfrac{x'^2}{4} + \dfrac{\sqrt{3}y'x'}{2} + \dfrac{3y'^2}{4} = 0$$

$$x'^2 + y'^2 + \sqrt{3}x' - y' = 0$$

15. $\cos \theta = .6$

$\theta = 53.13°$

17. $x = \dfrac{x'}{\sqrt{2}} - \dfrac{y'}{\sqrt{2}}$

$y = \dfrac{x'}{\sqrt{2}} + \dfrac{y'}{\sqrt{2}}$

$$3x^2 - 2xy + 3y^2 = 2$$

$$\dfrac{3(x'-y')^2}{2} - \dfrac{2(x'-y')(x'+y')}{\sqrt{2}\sqrt{2}} + \dfrac{3(x'+y')^2}{2} = 2$$

$$3(x'^2 - 2x'y' + y'^2) - 2(x'^2 - y'^2) + 3(x'^2 + 2x'y' + y'^2) = 4$$

$$4x'^2 + 8y'^2 = 4$$

$$x'^2 + 2y'^2 = 1$$

19. $x^2 + xy = 1$

$$(x' \cos \theta - y' \sin \theta)^2 + (x' \cos \theta - y' \sin \theta)(x' \sin \theta + y' \cos \theta) = 1$$

$$x'^2(\cos^2 \theta + \sin \theta \cos \theta) + y'^2(\sin^2 \theta - \cos \theta \sin \theta) +$$

$$x'y'(-2 \cos \theta \sin \theta - \sin^2 \theta + \cos^2 \theta) = 1$$

Let $-2 \cos \theta \sin \theta - \sin^2 \theta + \cos^2 \theta = 0$

$$\cos^2 \theta - (1 - \cos^2 \theta) = 2 \cos \theta \sqrt{1 - \cos^2 \theta}$$

$$(2 \cos^2 \theta - 1)^2 = \left(2 \cos \theta \sqrt{1 - \cos^2 \theta}\right)^2$$

$$4 \cos^4 \theta - 4 \cos^2 \theta + 1 = 4 \cos^2 \theta (1 - \cos^2 \theta)$$

$$8 \cos^4 \theta - 8 \cos^2 \theta + 1 = 0$$

$$\cos^2 \theta = \frac{8 \pm \sqrt{64 - 32}}{16} = \frac{2 \pm \sqrt{2}}{4}$$

$$\cos \theta = .9238, \quad \theta = 22.5°$$

$$x'^2(.8534 + .3535) + x'y'(-.707 - .1465 + .8534) + y'^2(.1465 - .3535) =$$

$$\frac{x'^2}{.828} - \frac{y'^2}{4.828} = 1$$

21. $x^2 - xy + y^2 = 1$

$$(x' \cos \theta - y' \sin \theta)^2 - (x' \cos \theta - y' \sin \theta)(x' \sin \theta + y' \cos \theta) +$$

$$(x' \sin \theta + y' \cos \theta)^2 = 1$$

$$x'^2(\cos^2 \theta - \cos \theta \sin \theta + \sin^2 \theta) + y'^2(\sin^2 \theta + \sin \theta \cos \theta + \sin^2 \theta) +$$

$$x'y'(\sin^2 \theta - \cos^2 \theta) = 1$$

Let $\sin^2 \theta - \cos^2 \theta = 0$

$$\sin^2 \theta = \cos^2 \theta$$

$$\theta = 45°$$

$$x'^2\left(1 - \tfrac{1}{2}\right) + y'^2\left(1 + \tfrac{1}{2}\right) = 1$$

$$\frac{x'^2}{2} + \frac{y'^2}{\frac{2}{3}} = 1$$

1. $x^2 + y^2 + 4x + 6y + 4 = 0$

   $x^2 + 4x + y^2 + 6y + 9 = -4 + 4 + 9$

   $(x+2)^2 + (y+3)^2 = 9$

   center: $(-2, -3)$,   radius: 3

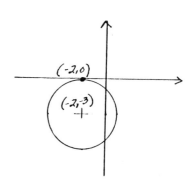

3. $9x^2 + 4y^2 + 18x + 8y - 23 = 0$

   $9(x^2+2x+1) + 4(y^2+2y+1) = 23 + 9 + 4$

   $9(x+1)^2 + 4(y+1)^2 = 36$

   $\dfrac{(x+1)^2}{4} + \dfrac{(y+1)^2}{9} = 1$

   center: $(-1, -1)$

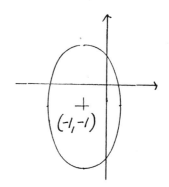

5. $x^2 - y^2 - 4x - 21 = 0$

   $x^2 - 4x + 4 - y = 21 + 4$

   $(x-2)^2 - y^2 = 25$

   center: $(2, 0)$,   radius: 5

7. $x^2 - 6x - 3y = 0$

$x^2 - 6x + 9 = 3y + 9$

$(x-3)^2 = 3(y+3)$

$4a = 3, \quad a = \frac{3}{4}$

Vertex: $(3, -3)$

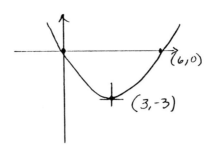

9. $2y^2 - 2y + x - 1 = 0$

$2\left(y^2 - y + \frac{1}{4}\right) = -x + 1 + \frac{1}{2}$

$2\left(y - \frac{1}{2}\right)^2 = -\left(x - \frac{3}{2}\right)$

$\left(y - \frac{1}{2}\right)^2 = -\frac{1}{2}\left(x - \frac{3}{2}\right)$

$4a = -\frac{1}{2}, \quad a = -\frac{1}{8}$

Vertex: $\left(\frac{3}{2}, \frac{1}{2}\right)$

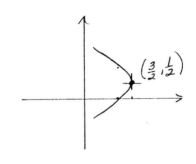

11. $\qquad 4x^2 - y^2 + 8x + 2y - 1 = 0$

$4(x^2 + 2x + 1) - (y^2 - 2y + 1) = 1 + 4 - 1$

$4(x + 1)^2 - (y-1)^2 = 4$

$(x + 1)^2 - \frac{(y - 1)^2}{4} = 1$

center: $(-1, 1)$

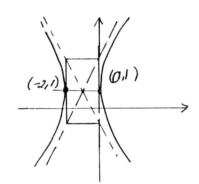

13. $9x^2 + 4y^2 - 18x + 16y - 11 = 0$

$$9(x^2 - 2x + 1) + 4(y^2 + 4y + 4) = 11 + 9 + 16$$

$$9(x - 1)^2 + 4(y + 2)^2 = 36$$

$$\frac{(x - 1)^2}{4} + \frac{(y + 2)^2}{9} = 1$$

center: $(1, -2)$

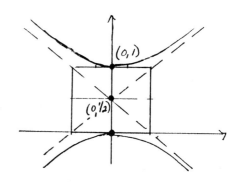

15. $x^2 - 2y^2 + 2y = 0$

$$x^2 - 2\left(y^2 - y + \tfrac{1}{4}\right) = -\tfrac{1}{2}$$

$$x^2 - 2\left(y - \tfrac{1}{2}\right)^2 = -\tfrac{1}{2}$$

$$-2x^2 + 4\left(y - \tfrac{1}{2}\right)^2 = 1$$

$$-\frac{x^2}{\tfrac{1}{2}} + \frac{\left(y - \tfrac{1}{2}\right)^2}{\tfrac{1}{4}} = 1$$

center: $\left(0, \tfrac{1}{2}\right)$

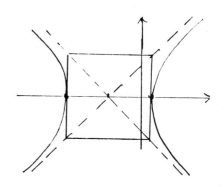

17. $x(x + 4) = y^2 + 3$

$$x^2 + 4x - y^2 = 3$$

$$x^2 + 4x + 4 - y^2 = 7$$

$$\frac{(x + 2)^2}{7} - \frac{y^2}{7} = 1$$

center: $(-2, 0)$

19.
$$y = x^2 + 5x + 7$$
$$y - 7 = x^2 + 5x$$
$$y - 7 + \frac{25}{4} = x^2 + 5x + \frac{25}{4}$$
$$y - \frac{3}{4} = \left(x + \frac{5}{2}\right)^2$$
$$4a = 1, \quad a = \frac{1}{4}$$
vertex: $\left(-\frac{5}{2}, \frac{3}{4}\right)$

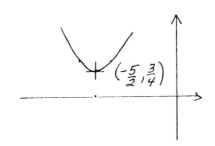

21. $5x^2 + 12xy = 9$

$A = 5, \quad B = 12, \quad C = 0$

$\cot 2\theta = \frac{5-0}{12} = \frac{5}{12}$

$\cos 2\theta = \frac{5}{13}$

$\sin \theta = \sqrt{\frac{1 - \frac{5}{13}}{2}} = \frac{2}{\sqrt{13}}$

$\cos \theta = \sqrt{\frac{1 + \frac{5}{13}}{2}} = \frac{3}{\sqrt{13}}, \theta = 33.69°$

$x = x'\frac{3}{\sqrt{13}} - y'\frac{2}{\sqrt{13}}$

$y = x'\frac{2}{\sqrt{13}} + y'\frac{3}{\sqrt{13}}$

$$5\left(\frac{3x' - 2y'}{\sqrt{13}}\right)^2 + 12\left(\frac{3x' - 2y'}{\sqrt{13}}\right)\left(\frac{2x' + 3y'}{\sqrt{13}}\right) = 9$$

$$5(9x'^2 - 6x'y' + 4y'^2) + 12(6x'^2 + 9x'y' - 4x'y' - 6y'^2) = 117$$

$$117x'^2 - 52y'^2 = 117$$

$$x'^2 - \frac{y'^2}{\frac{9}{4}} = 1$$

23. $5x^2 + 24xy - 2y^2 = 44$

$A = 5, \quad B = 24, \quad C = -2$

$\cot 2\theta = \dfrac{5 - (-2)}{24} = \dfrac{7}{24}$

$\cos 2\theta = \dfrac{7}{25}$

$\sin \theta = \sqrt{\dfrac{1 - \dfrac{7}{25}}{2}} = \dfrac{3}{5}$

$\cos \theta = \sqrt{\dfrac{1 + \dfrac{7}{25}}{2}} = \dfrac{4}{5}, \quad \theta = 36.87°$

$x = \dfrac{4}{5}x' - \dfrac{3}{5}y'$

$y = \dfrac{3}{5}x' + \dfrac{4}{5}y'$

$$5\left(\frac{4x' - 3y'}{5}\right)^2 + 24\left(\frac{4x' - 3y'}{5}\right)\left(\frac{4x' + 3y'}{5}\right) - 2\left(\frac{3x' + 4y'}{5}\right)^2 = 44$$

$$\frac{5}{25}(16x'^2 - 24x'y' + 9y'^2) + \frac{24}{25}(12x'^2 - 9x'y + 16x'y' - 12y'^2)$$

$$- \frac{2}{25}(9x'^2 + 24x'y' + 16y'^2) = 44$$

$14x'^2 - 11y'^2 = 44$

$\dfrac{x'^2}{\frac{22}{7}} - \dfrac{y'^2}{4} = 1$

25. $c^2x^2 - 8c^2x + 4y^2 + 12c^2 = 0$

$c^2(x^2 - 8x) + 4y^2 = -12c^2$

$c^2(x^2 - 8x + 16) + 4y^2 = -12c^2 + 16c^2$

$c^2(x - 4)^2 + 4y^2 = 4c^2$

$\dfrac{(x - 4)^2}{4} + \dfrac{y^2}{c^2} = 1, \quad$ ellipses

27. $y^2 + x^2 = (x + 5)^2$

$\quad y^2 + x^2 = x^2 + 10x + 25$

$\qquad y^2 = 10x + 25, \quad$ parabola

Section 14.7 Polar Equations and Their Graphs

11. $2x + 3y = 6$

$\quad 2(r \cos \theta) + 3(r \sin \theta) = 6$

$\quad r(2 \cos \theta + 3 \sin \theta) = 6$

13 $x^2 + y^2 - 4x = 0$

$\quad (r \cos \theta)^2 + (r \sin \theta)^2 - 4r \cos \theta = 0$

$\quad r^2 \cos^2 \theta + r^2 \sin^2 \theta - 4r \cos \theta = 0$

$\quad r^2(\cos^2 \theta + \sin^2 \theta) - 4r \cos \theta = 0$

$\qquad\qquad r^2 - 4r \cos \theta = 0$

$\qquad\qquad r(r - 4 \cos \theta) = 0$

$\qquad\qquad\qquad r = 4 \cos \theta$

15. $x^2 + 4y^2 = 4$

$\quad (r \cos \theta)^2 + 4(r \sin \theta)^2 = 4$

$\quad r^2 \cos^2 \theta + 4r^2 \sin^2 \theta = 4$

$\quad r^2(\cos^2 \theta + 4 \sin^2 \theta) = 4$

17. $x^2 = 4y$

$\quad (r \cos \theta)^2 = 4r \sin \theta$

$\quad r^2 \cos^2 \theta = 4r \sin \theta$

$\quad r \cos^2 \theta = 4 \sin \theta$

19. $r = 5$

$\quad \sqrt{x^2 + y^2} = 5$

$\quad x^2 + y^2 = 25$

21. $r = 10 \sin \theta$

$$\sqrt{x^2 + y^2} = \frac{10y}{\sqrt{x^2 + y^2}}$$

$$x^2 + y^2 = 10y$$

23. $r = 1 \pm 2 \sin \theta$

$$\sqrt{x^2 + y^2} = 1 + \frac{2y}{\sqrt{x^2 + y^2}}$$

$$x^2 + y^2 = \sqrt{x^2 + y^2} + 2y$$

25. $r = \dfrac{5}{1 + \cos \theta}$

$$\sqrt{x^2 + y^2} = \frac{5}{1 + \dfrac{x}{\sqrt{x^2 + y^2}}}$$

$$\sqrt{x^2 + y^2} = \frac{5\sqrt{x^2 + y^2}}{\sqrt{x^2 + y^2} + x}$$

$$1 = \frac{5}{\sqrt{x^2 + y^2} + x}$$

$$\sqrt{x^2 + y^2} + x = 5$$

$$x - 5 = \sqrt{x^2 + y^2}$$

$$(x - 5)^2 = x^2 + y^2$$

$$x^2 - 10x + 25 = x^2 + y^2$$

$$y^2 = 25 - 10x$$

31. $r = 2 \sin \theta$

θ	0	$\pi/6$	$\pi/4$	$\pi/3$	$\pi/2$	$3\pi/4$	π
r	0	1	$\dfrac{2}{\sqrt{2}}$	$\sqrt{3}$	2	$\dfrac{2}{\sqrt{2}}$	0

33. $r \sin \theta = 1$, $\quad r = \csc \theta$

θ	0	$\pi/6$	$\pi/4$	$\pi/3$	$\pi/2$	$3\pi/4$
θ	dne	2	$\sqrt{2}$	$\dfrac{2}{\sqrt{3}}$	1	$\sqrt{2}$

35. $r = 1 + \sin\theta$

θ	0	$\pi/6$	$\pi/3$	$\pi/2$	$2\pi/3$	$5\pi/6$	π
r	1	$\frac{3}{2}$	$\frac{\sqrt{3}+2}{2}$	2	$\frac{\sqrt{3}+2}{2}$	$\frac{3}{2}$	1

	$7\pi/6$	$4\pi/3$	$3\pi/2$	$5\pi/3$	$11\pi/6$	2π
	$\frac{1}{2}$	$\frac{2-\sqrt{3}}{2}$	0	$\frac{2-\sqrt{3}}{2}$	$\frac{1}{2}$	1

37. $r = \sec\theta$

θ	0	$\pi/6$	$\pi/4$	$\pi/3$	$\pi/2$	$2\pi/3$	$5\pi/6$	π
r	1	$\frac{2}{\sqrt{3}}$	$\sqrt{2}$	2	dne	-2	$\frac{-2}{\sqrt{3}}$	-1

	$7\pi/6$	$4\pi/3$	$3\pi/2$	$5\pi/3$	$11\pi/6$	2π
	$-\frac{2}{\sqrt{3}}$	-2	dne	2	$\frac{2}{\sqrt{3}}$	1

39. $r = 4\sin 3\theta$

θ	$\pi/6$	$\pi/4$	$\pi/3$	$\pi/2$	$2\pi/3$	$3\pi/4$	$5\pi/6$	π
r	4	$2\sqrt{2}$	0	-4	0	$2\sqrt{2}$	4	0

41. $r = 100(1 + \cos\theta)$

$r = 1 + \cos\theta$

θ	0	$\pi/6$	$\pi/4$	$\pi/3$	$\pi/2$	$2\pi/3$	$3\pi/4$	$5\pi/6$	π
r	1	$\frac{2+\sqrt{3}}{2}$	$\frac{2+\sqrt{2}}{2}$	$3/2$	$\frac{\theta}{r}$	$1/2$	$\frac{2-\sqrt{2}}{2}$	$\frac{2-\sqrt{3}}{2}$	0

	$7\pi/6$	$5\pi/4$	$4\pi/3$	$3\pi/2$	$5\pi/3$	$7\pi/4$	$11\pi/6$	2π
	$\frac{2-\sqrt{3}}{2}$	$\frac{2-\sqrt{2}}{2}$	$\frac{2-\sqrt{3}}{2}$	1	$3/2$	$\frac{2+\sqrt{2}}{2}$	$\frac{2+\sqrt{3}}{2}$	2

43. $r = \dfrac{1}{2 - \cos\theta}$

$$\sqrt{x^2 + y^2} = \dfrac{1}{2 - \dfrac{x}{\sqrt{x^2 + y^2}}}$$

$$\sqrt{x^2 + y^2} = \dfrac{\sqrt{x^2 + y^2}}{2\sqrt{x^2 + y^2} - x}$$

$$2\sqrt{x^2 + y^2} - x = 1$$

$$2\sqrt{x^2 + y^2} = x + 1$$

$$4(x^2 + y^2) = (x + 1)^2$$

$$4x^2 + 4y^2 = x^2 + 2x + 1$$

$$3x^2 + 4y^2 - 2x - 1 = 0$$

Chapter 14   Review

3.
$$x^2 + y^2 + 4y = 5$$
$$x^2 + y^2 + 4y + 4 = 5 + 4$$
$$x^2 + (y + 2)^2 = 3^2$$

5.
$$(x + 1)^2 - y^2 - 2y = 0$$
$$(x + 1)^2 - (y^2 + 2y + 1) = -1$$
$$(x + 1)^2 - (y + 1)^2 = -1$$
$$(y+1)^2 - (x+1)^2 = 1$$

7.
$$x^2 - y^2 = 100$$
$$\dfrac{x^2}{100} - \dfrac{y^2}{100} = 1$$

9.
$$y^2 + 2(x^2 + 3x) = 7$$
$$y^2 + 2\left(x^2 + 3x + \dfrac{9}{4}\right) = 7 + \dfrac{9}{2}$$
$$y^2 + 2\left(x + \dfrac{3}{2}\right)^2 = \dfrac{23}{2}$$
$$\dfrac{y^2}{\dfrac{23}{2}} + \dfrac{\left(x + \dfrac{3}{2}\right)^2}{\dfrac{23}{4}} = 1$$

11. Vertex: $(2, 0)$, $a = -2$

Focus: $(0, 0)$

$$(y - 0)^2 = 4(-2)(x - 2)$$
$$y^2 = -8(x - 2)$$

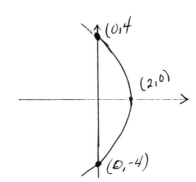

13. Center: $(-1, 2)$

Radius: 8

$$(x + 1)^2 + (y - 2)^2 = 64$$

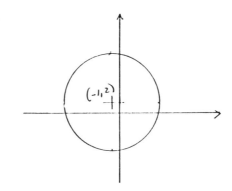

15. Center: $(0, 0)$

Focus: $(2, 0)$, $c = 2$

Major Axis: 8, $a = 4$

$$b^2 = 16 - 4 = 12$$
$$\frac{x^2}{16} + \frac{y^2}{12} = 1$$

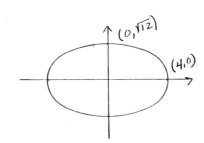

17. Vertices: $(7, 4), (-1, 4)$

Asymptote: $y = \pm 2x$

Center: $(3, 4)$

$a = 4, \quad b = 8$

$\dfrac{(x - 3)^2}{16} - \dfrac{(y - 4)^2}{64} = 1$

19. Major Axis: $30, \quad a = 15$

$b = 12$

$\dfrac{x^2}{15^2} + \dfrac{y^2}{12^2} = 1$

$c^2 = 15^2 - 12^2 = 81, \quad$ Foci: $(\pm 9, 0)$

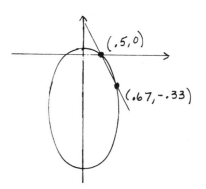

21.
$$2x^2 + x + y^2 + 2y = 1$$
$$2\left(x^2 + \tfrac{x}{2}\right) + y^2 + 2y = 1$$
$$2\left(x^2 + \tfrac{x}{2} + \tfrac{1}{16}\right) + (y^2 + 2y + 1) = 1 + \tfrac{1}{8} + 1$$
$$2\left(x + \tfrac{1}{4}\right)^2 + (y + 1)^2 = \tfrac{17}{8}$$
$$\dfrac{\left(x + \tfrac{1}{4}\right)^2}{\tfrac{17}{16}} + \dfrac{(y + 1)^2}{\tfrac{17}{8}} = 1$$

$y + 2x = 1, \quad y = -2x + 1$

23. See p. 447 and Fig. 14.1 in the text

25. $y = 2x$

$\frac{y}{x} = 2$

$\tan \theta = 2$

29. $4(r \cos \theta)^2 + (r \sin \theta)^2 = 1$

$4r^2 \cos^2 \theta + r^2 \sin^2 \theta = 1$

31. $r = 2$

$\sqrt{x^2 + y^2} = 2$

$x^2 + y^2 = 4$

33. $r = \dfrac{2}{1 + \sin \theta}$

$$\sqrt{x^2 + y^2} = \dfrac{2}{1 + \dfrac{y}{\sqrt{x^2 + y^2}}}$$

$$\sqrt{x^2 + y^2} = \dfrac{2\sqrt{x^2 + y^2}}{\sqrt{x^2 + y^2} + y}$$

$$\sqrt{x^2 + y^2} + y = 2$$

$$y - 2 = -\sqrt{x^2 + y^2}$$

$$(y - 2)^2 = x^2 + y^2$$

$$y^2 - 4y + 4 = x^2 + y^2$$

$$x^2 = -4(y - 1)$$

35. $r = 3 \cos \theta$

$$\sqrt{x^2 + y^2} = \dfrac{3y}{\sqrt{x^2 + y^2}}$$

$$x^2 + y^2 = 3y$$

$$x^2 + y^2 - 3y + \dfrac{9}{4} = \dfrac{9}{4}$$

$$x^2 + \left(y - \dfrac{3}{2}\right)^2 = \left(\dfrac{3}{2}\right)^2$$

37. $r = \sin 2\theta$

$r = 2 \sin \theta \cos \theta$

$$\sqrt{x^2 + y^2} = 2\left(\frac{y}{\sqrt{x^2 + y^2}}\right)\left(\frac{x}{\sqrt{x^2 + y^2}}\right)$$

$(x^2 + y^2)^{3/2} = 2xy$

39. $r^2 - r = 0$

$r(r - 1) = 0$

$r = 1$

$\sqrt{x^2 + y^2} = 1$

$x^2 + y^2 = 1$